FAMOUS SCIENTIFIC EXPEDITIONS

By RAYMOND B. HOLDEN

Have you ever wondered what lies on the other side of a locked door? This same sense of wonder is the spur that urges explorers to batter down the walls between us and a greater understanding of our world.

In this book, the author gives dramatic accounts of five explorations that opened doors to new corners of knowledge. Two of the stories are built around the adventures of men who turned back the pages of history and brought the past to life. Howard Carter's search for the tomb of the Pharaoh, Tut-Ankh-Amen, was an effort to discover how the Egyptians lived thirty centuries ago. A story of a vastly different adventure tells about the Burden expedition to a primitive Pacific island to hunt the "dragons" of fairy tales. It had just been learned that the fantastic beasts, which supposedly had died out millions of years before, were still living on Komodo Island!

Two of the adventures were danger-filled expeditions to extreme heights and depths of the earth. The conquest of Mt. McKinley by Dr. Hudson Stuck and his party opened a new era in mountain climbing. And when William Beebe and Otis Barton climbed into their bathysphere and plunged into the ocean, they became the first men to witness the lives of creatures that dwell deep below the surface of the sea.

In the last story we meet a man of extraordinary courage—an Arctic explorer. Vilhjalmur Stefansson was determined to open the way to Arctic travel—if only he could find ways to cope with the Arctic's mischievous tricks.

These heroic exploits are proof that new frontiers await the adventurer who longs to broaden the limits of knowledge. Those who do their adventuring at home will find in these pages five tales that carry the reader to extraordinary places and open his eyes to unusual scenes.

Illustrated by LEE J. AMES

RANDOM HOUSE • NEW YORK

journal which reflects so clearly the multiplicity of his interests, and which may one day be the basis for his own "Natural History of New Hampshire."

Famous Scientific Expeditions

Famous Scientific Expeditions

by Raymond Holden

Illustrated by Lee Ames

Random House　　　New York

I wish to express my indebtedness to those men of science who have permitted me to use their own accounts of their expeditions as source material. To Dr. William Beebe, for material from his book, *Half Mile Down*; to Mr. W. Douglas Burden for permission to quote from his volume, *Dragon Lizards of Komodo*; and to Dr. Vilhjalmur Stefansson for his generous help in adapting the story which he himself has told in *The Friendly Arctic*.

For permission to use material from books published by them I have to thank the following: The Macmillan Company, publishers of *The Friendly Arctic*, by Vilhjalmur Stefansson, copyright 1921 by The Macmillan Company, and 1949 by Vilhjalmur Stefansson; Duell, Sloane & Pearce, publishers of *Half Mile Down*, copyright 1934 by William Beebe; G. P. Putnam's Sons, publishers of *The Conquest of Mt. McKinley* by Belmore Browne, copyright 1913, and of *The Dragon Lizards of Komodo* by W. Douglas Burden, copyright 1927; Charles Scribner's Sons, publishers of *The Ascent of Denali* by Hudson Stuck; Cassell & Company, Ltd., of London, England, publishers of *The Tomb of Tut-Ankh-Amen* by Howard Carter.

RAYMOND HOLDEN

Contents

Foreword

Man is unusual among living creatures in that he can plan and devise ways to get what he wants. He likes to go places, so he learns how to make ships, trains, automobiles, and airplanes. During his travels he learns more and more about the world and what is in it.

But ships, trains, automobiles, and airplanes do not, by themselves, help man to solve all the world's problems or to pry open all the world's secrets. At any rate, he wants to know more than just *what*; he wants to know *why* and *how*. He is not satisfied to know himself as he is today; he also wants to know how he got to be that way. So he looks for a gateway to the past to show him what he was and finds it perhaps in the tomb of an ancient king in Egypt.

Man's curiosity extends to the other living creatures who share the earth with him. The gateway to this knowledge he may find in the rain forests of Africa, or in the jungles of New Guinea. He looks at a mountain, so high that its peak seems beyond his reach. Nevertheless, curious man wants to know what it is like up there, and so he is off on another adventure.

This book will try to tell you the story of some of man's trips through the gateways to knowledge—trips which have taken him into deserts and jungles, to the tops of mountains, and to the depths of the sea.

Famous Scientific Expeditions

1. Journey to the Depths of the Sea

The Beebe-Barton Bathysphere

When ancient man stood on the shore and looked seaward, he saw what seemed to be nothing but water. An enormous flat plain of blue-gray water! Today we know that the vast sea is not so simply described—it is a world in itself, a world full of life, full of wonders and mysteries.

We human beings have always been surrounded by the sea. The ancients thought of it as a great river coiled like a snake about the world's land. Indeed, although the sea does not flow in a riverlike stream, it does surround the land, covering the greater part of the globe.

From the very earliest days men must have wondered if there was any end to the great water, but they did not just keep on wondering. They found out. They learned how to make ships and how to sail them. They learned to use a compass and to study the stars in order to know in what direction they were traveling.

Slowly they discovered what was on the other side of the water and almost all there was to know about its surface. Yet the sea was still a mystery, for it was not only wide, it was very, very deep. In its greatest depth the bottom of the sea is farther below the surface than the highest mountains on land are above it. At one spot near Japan the ocean is almost 35,000 feet or six and one-half miles deep. The average depth is more than two miles.

When you realize that until 1930 no living person had ever been able to plunge much more than 500 feet below the surface, you can understand why the sea had never been thoroughly explored. Let us examine some of the reasons for this.

First, as everybody knows who has ever tried to swim, we cannot breathe water. If we try to do so, we drown. Nor can we stop breathing for long enough to get very far down.

4

Then, too, water is very heavy, much heavier than air, as anyone who has ever tried to carry a pail of it knows. The deeper you go, the more water is piled on top of you. At great depths, the sea is very cold and very dark. In shallow pools you can sometimes see the sun shining on the bottom and may, therefore, think that light passes through water as it does through a window glass. This is not so. The sun cannot penetrate very deeply into the sea, and in the deepest place there is no light at all.

Yet as long ago as the time of Alexander the Great, more than 300 years before Christ, men were thinking about how they might reach the bottom of the ocean to see what was there. They were curious because from earliest times there have always been tales of strange sea creatures that had been found on the shore or tangled in fishermen's nets. Perhaps, folks thought, there were still stranger and more wonderful things hidden in the cold black deep!

Curiosity was not the only cause of the longing to explore the ocean floor. Some men wanted to go there just for the excitement of being where no one had ever been before and seeing what no one had ever seen. Some wanted to find and bring up the rich treasures of sunken ships. Others wanted to find out something about the secret of life which may have begun in the depths of the sea.

Long ago men began to try to find ways of plunging deep into the water. If only they could breathe under

5

water like the fish and the other sea creatures! Since this was not possible, man's attempts to search the depths of the sea have had to be made with the help of air taken down into the water.

You know that if you cup your hands when you are in the water and then turn them over and press them down, you can carry air below the surface. The earliest divers knew this, so they used such things as pails or tubs. When these objects were turned upside down under water, they would hold air. By sticking their heads up into this air, divers were able to breathe. This plan, which worked in shallow water, is the principle used in today's diving helmets.

There was, however, a great deal of trouble in staying down. The more light air a man carried, the more weight he had to attach to himself to stay under water. The invention of the diving suit, into which air could be pumped from above, solved this problem for depths up to 500 feet. Even at that depth, however, the suit was so heavy and clumsy that a man who wore one found it almost impossible to do anything once he hit bottom.

So far, nothing has been said about submarines. It might seem that because they can carry men and move about below the waves under their own power, they would be the answer to the problem. After all, in Jules Verne's story, Captain Nemo traveled 20,000 leagues under the sea in the submarine *Nautilus*! Readers of this tale might say, "All that is needed to explore the depths of the ocean is to perfect the undersea boat." However, although submarines have made great progress, no submarine has yet been able to operate safely at more than 500 feet below the surface.

In the 1920s Dr. William Beebe, Director of the Tropical Research Department of the New York Zoological Society, joined the ranks of those who dreamed of visiting the sea's depths. Dr. Beebe had spent much time sailing the ocean, scooping up its life in nets and deep trawls, finding fish and other forms of life which had been entirely unknown, and learning new facts about the familiar varieties. Rich as his collections were, he knew that he was missing more than he had found. He kept dreaming of a personal journey to the depths of the ocean, so that he might see its life with his own eyes.

To accomplish this he planned several metal diving cylinders, but he soon realized that they would not be able to stand the fearful pressure of great depth. For every 33 feet of dive, an additional weight of nearly 15 pounds would have to be added to every square inch of the cylinder's outer surface. Half a mile down, there would be more than a ton pressing against every inch. The problem seemed to be one that would never be solved.

Yet it was solved—and by a man named Barton who

later became associated with Dr. Beebe. Otis Barton
was interested in under-water photography and he, too,
wanted a way of getting farther down into the sea than
a diver's helmet or suit would carry him. By 1930, he
had been able to put up enough money and provide
enough ideas to interest a firm of naval architects. With
Captain John Butler he had worked out what he believed
would be a practical way of exploring the ocean. Barton's idea was based on the fact that a ball-shaped
object will stand more pressure than any other shape.
Beebe and Barton decided to work together on an attempt to do what no man had yet been able to do.

Who would believe that a hollow steel ball less than
five feet in diameter, with walls only an inch and a
quarter thick, could safely carry two men into the un-
explored depths of the sea? Many people, including
Beebe and Barton themselves, were somewhat doubtful
that it could be done, but the two explorers were determined to find out. First, Dr. Beebe wanted a name
for the strange device with its three great eyes that
stuck out like a crab's. He decided to add to the word
"sphere" the Greek word for depth, which made the
name *bathysphere*.

Beebe took a staff of scientists and mechanics with
him to Nonsuch Island, Bermuda. There the New York
Zoological Society had a research laboratory, a place
where the problems of life in the sea are studied.

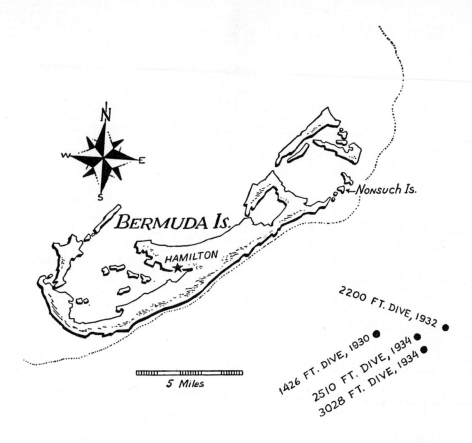

A great deal of equipment was needed for the adventure. Dr. Beebe took along a sea-going tug, an open-decked barge equipped with a derrick and steam engine, and a huge reel to carry the cable by which the bathysphere would be lowered into the sea.

Barton provided the sphere itself and more than half a mile of steel cable that was seven-eighths of an inch

9

The barge Ready *off Nonsuch Island, Bermuda*

thick. He also contributed a solid rubber cable in which were embedded two sets of electrical wires. One wire was for electric light and one was for a telephone. These were necessary, because the men in the bathysphere would have to talk to their helpers up above. Electricity would also be needed to light up the dark sea outside the sphere.

Let us take a look at the deck of the barge *Ready* as she rode the calm swells eight miles off Nonsuch

10

Island, Bermuda, one morning in June, 1930. The old wooden flat-deck is held into the wind and slow-rolling swells by a steam-tug while the expedition staff—nineteen persons in addition to Dr. Beebe and Mr. Barton —bustles about getting things ready.

In everyone's mind there is a feeling of excitement. Each has his own duty to perform. One keeps an eye on the steel cable, while several more take care of the rubber cable that contains the electric wires. Two others work the derrick winches and watch the boilers in the hoisting engine. Still others, who will have more to do when the actual descent has begun, stand by to give a helping hand.

Barton and Dr. Beebe go into the bathysphere to see that all is well. Their task is not a comfortable one, for the hole by which they enter the sphere is only 14 inches in width and is ringed with great protruding bolts. Anyone entering the sphere has to crawl headfirst over these bolts and into the hole.

Under the barge there is more than a mile of water. Neither explorers nor crew doubt that the diving sphere can be lowered as far as the length of the cable will allow. Yet no one can be sure the two divers will be alive when the sphere is hauled up again. Perhaps the bathysphere will not be able to stand the enormous pressure of the great depths. Perhaps the cold black water will be forced in around the edges of the windows and

11

door. Perhaps it will seep through where the electric cable enters the steel shell. A joint which will not leak just under the sea's surface might well give up under the fearful pressure of six and a half million pounds a quarter of a mile down. These explorers are scientists, not daredevils. They must know the answer to every question before they risk their lives.

So, on June 3, 1930, the monster with its crab-eyes is sealed up, with no one in it, and swung out over the water. Down, down it goes, slowly and steadily. For 75 or 100 feet it is visible in the clear water. Then it is lost in the darkness. In three quarters of an hour 2,000 feet of cable have been reeled out. Now the bathysphere hangs suspended almost halfway to the bottom! If anything has happened to the huge diving ball, it will have happened by this time. Dr. Beebe gives the signal to haul in.

As the cable is reeled in, he sees that something has gone wrong. The rubber cable containing the electric wires, which was fastened to the steel cable by which the sphere was lowered, has wound itself around the steel. As a result, the lifting wire cannot be reeled in without crushing the rubber cable and its telephone and light wires. Without this means of getting electricity down to the sphere no further dives can be attempted. There is nothing to do but to let the rubber cable slip down in a great coil (half a mile in length) about the sphere itself.

12

Beebe's heart sank. He began to wonder if man was meant to venture so far beneath the sea. Fortunately, after 24 weary hours of work, the rubber cable was untangled and laid out in large loops on the barge deck. Tests showed that the electric system had not been damaged.

Three days later the bright Bermuda sun looked down upon another experimental dive—this time to a depth of 1500 feet. All went well. The cables behaved perfectly and only a little water seeped in—barely a coffeepot full.

Now there was no reason for delay. If an empty sphere could go so deep, so could the men who had made the great diving ball.

Beebe and Barton climbed in, wincing at the chill of the steel floor. They folded themselves up as well as they could in the little round chamber. It was something like putting two men in a bathtub and turning another bathtub upside down over it.

They took along two tanks of oxygen and trays of chemicals to absorb the moisture of breath and the carbon dioxide that men breathe out. The oxygen, which fizzed very slowly from the tank, replaced what the two divers used up of the air in the sphere. It was necessary because there was no air pipe connected with the surface as there is in an ordinary diving suit or helmet.

The explorers soon settled themselves, with Beebe at one of the sphere's crab-eyes and Barton where he could

13

The crew began to tighten the huge bolts

watch the instruments. Outside, the crew began to tighten the huge bolts on the 400-pound steel door. To get them absolutely tight the wrenches had to be hit with sledge-hammers. You can imagine what this sounded like inside the sealed steel ball! Dr. Beebe says it was the most infernal racket he ever heard. When he had stopped protesting over the telephone all was ready.

The winch turned slowly and the two-and-a-half-ton steel ball with its two squatting passengers rose into the air and swung out over the sea. At a signal from the deck captain the winch was reversed and the sphere dropped toward the water. It struck with a splash and sank in a green burst of bubbles and spray.

Beebe was at the window that was made of quartz three inches thick. Above him he saw the bottom of the barge slowly disappearing. At last it could be seen no more and the world vanished with it. Now there was nothing but space, as truly as if the divers had been in a rocket ship going up instead of down. But this space was full of water. Down, down, down. Fifty feet! One hundred feet! Two hundred!

At 300 feet Barton let out a yell. He had spied a little trickle of water coming in under the bolted door. Beebe studied it anxiously. The flow did not increase, but he knew that the pressure which was causing it would grow greater with every foot of descent. There was nothing to do but go on down and see what happened. On

Dr. Beebe at the window of the bathysphere

they went. Outside the quartz window, green turned
to blue.

Soon the greatest depth ever reached by a diver was
passed. The real exploration had begun. From now on
they were to see a world that no one else had ever seen.

A trickle of water was coming in under the door

The blueness of the water outside changed slowly to black.

Every minute or two Beebe or Barton turned a flashlight on the leak under the door. Was it really not getting worse or did they just want to think so? At 800

feet Beebe gave the order over his telephone to stop
the descent. He sat in darkness for a while, thinking.
They were comfortable, the air was good and the leak
did not seem to be getting any bigger. Then suddenly,
and for no reason, he had a strong feeling that they
must not go on. He gave the signal to rise.

They learned what might have happened if they had
decided to go deeper, when, later, a practice dive was
made with an empty sphere. On this occasion the bathysphere was sealed up, swung over the side and dropped
3,000 feet. After an hour and 40 minutes it was hauled
up.

As the sphere hung above the deck ready to be lowered and inspected, Dr. Beebe saw at once that something was wrong. A sharp stream of water was shooting
across the face of one of the windows. When the apparatus settled to the deck Beebe saw that it was nearly
full of water.

This is what had happened. Before its dive, the sphere
had been filled with air. The water that had forced its
way into the ball had pressed the air into a thin layer at
the top of the round chamber. It was very angry air.
It wanted to swell to its original size, and it was fight-
ing with the water that had taken its place.

Dr. Beebe could see the water trembling as the air
struggled to get out through the tiny leak by which

the sea had come in. He began to unscrew the center bolt of the steel door. The furious air inside began to whine and shriek and to blast the intruding water out through the bolt hole in a cloud of hissing spray.

Realizing what might happen, Beebe ordered everyone away from the sphere. Then he and a helper turned the brass handles of the center bolt. More and more screaming water mixed with air rushed out. Suddenly the great brass bolt was torn from their hands and hurled across the deck like a cannon shot. A shaft of water hard as steel followed it. Had Beebe been in the way of either the bolt or shaft of water as they burst from the sphere he would have been torn to pieces.

Every man was glad that he had not been in that steel ball 3,000 feet down when the trapped air and the deep water began their struggle. The trouble had been caused by a newly inserted quartz window which had not been properly embedded in its frame.

Four days later, the empty sphere was dropped to 2,000 feet. This time there were no troubles, so Beebe and Barton climbed into their steel ball and once more headed for the unknown. At a depth of 250 feet, the telephone gave out. Without voices from the upper world, the two divers were dreadfully alone and helpless, but fortunately the electric light was still working. By blinking it in a prearranged signal they were able to make

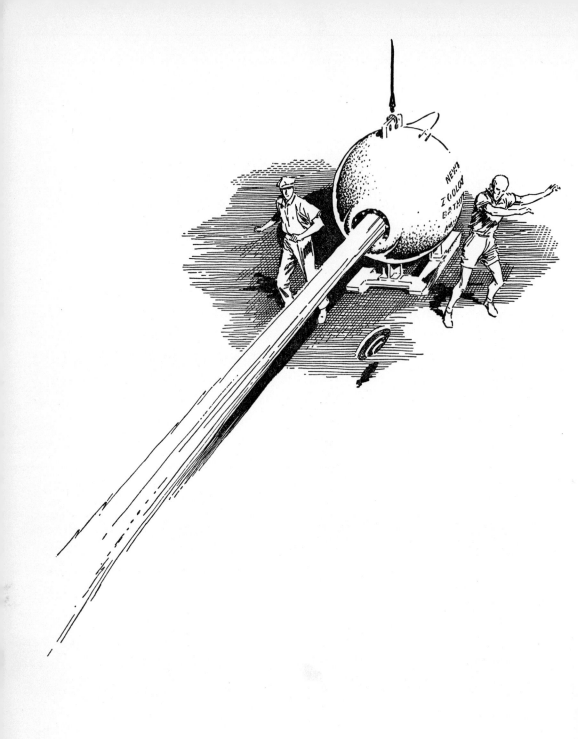

a light on the barge go on and off. This told the surface crew that Beebe and Barton wanted to be hauled up.

On July 11 the weather was perfect. The explorers took off for the depths without first sending the sphere down empty. Beebe felt that they now knew enough about the way it worked. He began to think less about their chances of getting down and up again and more about looking out the window for signs of life. Heretofore he had seen nothing that he could not have seen from the surface. Now he and his companion became explorers.

At 200 feet on this dive, Beebe saw a fish which he recognized—a pilot fish, common in Bermuda waters. But it should have been blue, deep blue with paler bands. This one was not. It was ghostly white, with black bands. Here was another indication of the difference between the land world and the undersea world. Even colors were different! You can have colors only where there is light as we know it.

At 400 feet the first deep sea fish, fish which are never seen alive at the surface, swam before the window. The world of the sea became a realm of mystery and magic, full of flashing sparks. Fish with streamers of lights along their sides! Creatures that could be seen through, with stomachs full of glowing embers! Golden snails without shells that flapped their wings and flew!

21

Suddenly the great brass bolt was torn from their hands and hurled across the deck like a cannon shot.

Silver hatchet fish, alive with ghostly radiance! Shrimps that blew out clouds of light as if they were water-pistols filled with liquid fire!

Slowly and steadily the bathysphere descended. At a little over 1,000 feet, the strong searchlight that was shining out of one of the quartz windows showed strange fish with greenish lights on their sides. Neither Dr. Beebe nor anyone else had ever before seen such creatures.

At 1,300 feet, or a third of a mile down, the sea was like a night full of shooting stars and bursting rockets. Sometimes unrecognized creatures swam against the window and exploded in a shower of sparks.

In the darkness, almost everything carried its own light which could be turned on or off at will. Almost, but not quite everything—for sometimes, peering out along the beam of his searchlight, Dr. Beebe could see the dim outlines of huge dark creatures moving away from the light. What were they? Some day we shall know. The mysteries of the deep are too many to be solved in a few short dives.

At 1,426 feet the bathysphere stopped. Here, Dr. Beebe says, he watched a fantastic little fish called a dragonfish, no more than six inches long, swimming easily in the light from the bathysphere. It seemed impossible for him to believe that he could not open the

23

Huge dark creatures moved away from the light

door of the sphere and swim out as easily as that little fish. Yet if he had tried to do so he would not even have drowned. The great pressure would have driven the first drops of water inward with such force that they would have gone through him like bullets.

The two divers went no farther that day, but twice in the following four years they tried again in the same tough little ball of steel. On the last dive, in August, 1934, they reached a depth of 3,028 feet. At this level they saw even larger and stranger creatures than they had found in the layers of water higher up. The ancestors of some of these had lived in the depths of the ocean perhaps millions of years before there was any such creature as man.

Although Dr. Beebe's pioneer dives in Otis Barton's bathysphere opened the way to an unvisited region, there are tens of thousands more feet in the ocean's depth which remain to be explored. The bathysphere, like the first locomotive or the first airplane, will seem crude and perhaps laughable to the scientist-explorers of the future. Still, that first bathysphere made it possible for man to visit unknown worlds.

Recently Dr. Auguste Piccard, a Belgian scientist, and officers of the French Navy have more than once descended to a depth of over 10,000 feet in the Mediterranean Sea! We may learn how whales, with no bathysphere to help them, can dive thousands of feet while men cannot. We may learn of sea-serpents living in the black caverns of the great water.

We may learn how the living beings of the depths can make and control light to guide them through the darkness.

One more great gateway to knowledge is open. It is only a question of time before we may all walk through it.

2. On Earth in the Upper Air

The Conquest of Mt. McKinley

The highest mountain in Asia—and in the world—has at last been climbed. After years of struggle, men have finally done what once seemed impossible. Yet although Mt. Everest will not be forgotten, now that the heroic story of its conquest has been told it is unlikely that climbers will pay much more attention to it.

Yet all mountains do not die once they have been conquered by men. There is one huge peak which has now been climbed more than half a dozen times and yet remains one of the most fascinating as well as the most beautiful of mountains. Strangely enough it was neither named nor located on the map until 40 years after Mt. Everest had been named and measured.

Those who do not know mountains from climbing experiences may be pardoned for thinking that a mountain is a mountain and that's that. Nothing could be further from the truth, as all climbers know. It is true that Mt. Everest is the highest mountain in the world and has proved to be one of the most difficult to climb. Yet there are lower peaks which have not been climbed at all, in spite of the mountaineer's art.

There are mountains more than 20,000 feet high in South America which may be climbed without once setting foot on ice or snow. On the other hand, there are mountains of half that height in North America which are so difficult that they can be conquered by only the hardiest and most experienced climbers.

Until a very few years before the beginning of the present century no one was sure which of the mountains north of Mexico was the highest. There had been rumors that in the heart of Alaska there existed a great peak which might possibly be the highest in the world. Yet it was so hard to reach that for years no one got close enough to it to find out.

In 1889 one man in search of gold got a view of the mountain and told of it with great enthusiasm on his return to civilization. In 1896 another man came close enough to the unnamed peak to make a very good guess as to its height and, in the opinion of many, a very poor guess as to what it ought to be called. He said it was 20,300 feet high and called it Mt. McKinley—after the not too distinguished gentleman who was a candidate for the office of President of the United States in that year.

Fortunate or not, the prospector's choice of a name stuck. Mt. McKinley was at last on the map, a little south of the exact center of Alaska. But it was seven years before anyone tried to climb it, and even then the attempt ended before a dent was made in its snows.

There was good reason for failure. The huge mountain, called "The Great One" by the Indians, stood in the center of a wilderness which had never been mapped. The nearest inhabited places, other than mining camps or trading posts, were Fairbanks, 150 miles northwest on the Tanana River, and Anchorage, 150 miles southeast on the shore of Cook Inlet. Between Mt. McKinley and Bering Sea lay nearly 400 miles of mountain, forest, river, and frozen marshland. To the east, more than 300 miles away, through more forest and across moun-

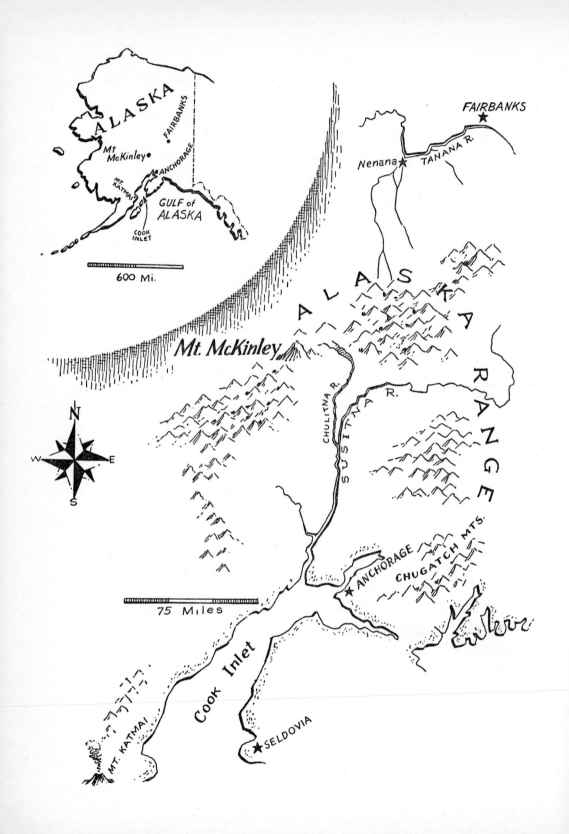

ALASKA

FAIRBANKS

Mt McKinley

ANCHORAGE

MT. KATMAI?

GULF of ALASKA

COOK INLET

600 Mi.

FAIRBANKS

Nenana

TANANA R.

ALASKA RANGE

Mt. McKinley

CHULITNA R.

SUSITNA R.

N

W   E

S

ANCHORAGE

CHUGATCH MTS.

75 Miles

Cook Inlet

MT. KATMAI

SELDOVIA

tain ranges and other great rivers, was the Canadian border.

The peak itself is perhaps the greatest single mountain mass in the world. Unlike Asia's Mt. Everest, which rises only a little more than 13,000 feet above its base at the mouth of the Rongbuk Glacier, Mt. McKinley towers at least 18,000 feet above the low foothills from which it rises. The only inhabitants of these hills are moose, caribou, bear, mountain sheep, and wolves. Also, unlike Mt. Everest, most of McKinley's 18,000 feet are permanently covered with snow and ice.

A mountain like McKinley simply cannot exist, even in a wilderness, without tempting men to climb to its top. Strangely enough, primitive people did not feel the challenge of a high mountain, or if they did, they were discouraged by fear. It was their belief that high places were the abode of savage demons and gods who would not care to be disturbed. Among civilized people there may also be some who never venture to climb the heights about their homes because they are perfectly certain that there is nothing whatever on the other side of them.

For the most part, however, men are curious and adventurous. This longing for adventure has given Mt. McKinley another distinction. It is so attractive to men that at least one of them tried to cover himself with

31

glory by saying that he had been to the top when he had not really been anywhere near it.

Mt. McKinley is like Mt. Everest in that before any attempt to climb it there had to be several expeditions for the purpose of finding out how to approach it.

The attack on Mt. McKinley began in earnest in 1906, although an attempt on it had been made in 1903. This early attempt was so unsuccessful that it barely touched the base of the great mountain mass and added nothing to man's knowledge of how to reach the summit.

In the middle of May, 1906, Professor Herschel Parker, Mr. Belmore Browne, and Dr. Frederic A. Cook joined forces in an expedition toward Mt. McKinley. They started out from Seldovia, Alaska, near the mouth of Cook Inlet 250 miles south of the majestic and mysterious mountain. The party was a large one, equipped with 20 pack horses and a motor launch, several packers, horse-handlers and helpers, a surveyor, and a photographer. This was a real expedition, with plenty of food and to all appearances plenty of know-how.

Professor Parker was an experienced mountaineer. So was Mr. Browne. Dr. Cook had been with Peary in the Arctic and had been through much of the wild country south of McKinley. From mid-May until August the expedition struggled with water, ice, snow, bogs, and mosquitoes. Although it did not get within striking distance of its goal, it proved that horses were unsuitable beasts of burden for such a trip. The discoveries made on the journey also seemed to make it plain that Mt. McKinley could not be climbed from the south. There was another outcome of this scouting trip which was one of the most fantastic things in the entire history of mountain climbing.

The amazing story began when the 1906 expedition broke up that summer. Instead of resting, Dr. Cook decided to go back up the Susitna River to see if there were any indications that the mountain which the northern Indians called Denali, or The Great One, could be reached by that route. Browne wanted to go along with him, but the Doctor insisted that he was merely going to look around and would not be doing any real exploring. Furthermore, he would rather have Browne do some hunting for him in the Chugatch Mountains. An eastern museum, said Cook, had asked him to get some specimens of Alaskan animals from that region.

Browne, somewhat reluctantly, went about his hunting and returned to Seldovia. There he heard the rumor, only a short time later, that Dr. Cook had climbed Mt. McKinley. Browne did not hesitate to say that such a rumor could not be true. He knew, as others familiar with the country did, that Dr. Cook had not had time to get to the mountain and back, much less climb it.

To Browne's amazement Dr. Cook and his companion soon appeared in Seldovia, and the Doctor admitted that he had been to the top of the mountain. Still doubtful of the tale, Browne took aside Dr. Cook's companion, Ed Barrill, who had been a good friend of his, and asked him what he knew about Mt. McKinley. Barrill would not say that he had been to the mountain. "Go and ask Cook," he said. After that Browne was sure. Yet even when the amazing Doctor published a book, *The Top of Our Continent*, including a photograph captioned "The Summit of Mt. McKinley," Browne's hands were tied.

Cook, who by that time also claimed to have reached the North Pole, refused to answer questions before a committee of The Explorers Club. His mind, he said, had been so unsettled by his grueling experiences that he could not speak without consulting his diary. He asked for two weeks to prepare himself. During the two weeks he disappeared, leaving the world divided between belief and disbelief.

This was the situation in 1910, more than 20 years after Mt. McKinley had first been described. Men were puzzled, but the giant peak stood as it had stood for thousands of years. Beautiful in white and gold, and mottled with purple ice-shadow, it seemed to rest on the earth as lightly as a cloud. Had man set foot on its summit or was it still untouched? Professor Herschel Parker and Belmore Browne believed, with much justification, that it had not been climbed. The great mountain still drew them and they knew that they must try again, if only to disprove Dr. Cook's claim.

Early in May, Parker and Browne set out with a well-equipped party, but this time they took no horses. Again they approached the great mountain from the south. If you look at a map, you will see that the vast Alaska range, of which Mt. McKinley is a part, stretches north and east like an unfolded wing that rises from the Alaskan Peninsula west of Cook Inlet. The entire region southeast of this great wing of mountains is a network of swift and stony rivers which empty into the Gulf of Alaska. Rushing through the swift streams are icy waters from the enormous glaciers of the great range.

It seemed obvious to Parker and Browne that by following the largest of the rivers to its source, a party could not help coming to the biggest glacier which, in all probability, would lead to the highest mountain slopes. For this purpose, the explorers chose the Susitna River, which flows into the north end of Cook Inlet. Then they followed its tributary, the Chulitna, to a point 40 miles southeast of the mountain where an enormous glacier poured a torrent of milky ice water into the river. Here, after nearly six weeks of traveling, they made their base camp and prepared for what they supposed would be the last stage of the conquest of McKinley.

Once more they were to learn that in the Alaskan Range things are seldom what they seem. The glacier up which they planned to travel looked like a broad snow highway that led gently toward the mountain. In reality it was a murderous tangle of icy ridges and deep crevasses. These were deceitfully covered with snow upon which the sun glared so strongly that the travelers were frequently forced to stop by the fierce pains of snow-blindness. Each time they halted, the explorers were beset by blizzard cold and the deep-drifting snow.

Although the climbers kept on until late in July they were unable to make much progress. Yet their efforts were rewarded. Halfway up the great glacier they saw a valley of ice between rocky snow-capped ridges. This point was about as far as Dr. Cook could have gone in the time he was away. The men believed that the photographs in his book, which he claimed to be pictures of the summit of Mt. McKinley, must have been taken near by. Climbing the ridge at the head of the valley, they soon saw that they were right.

Belmore Browne took a photograph that was exactly like the picture which Dr. Cook had called the summit of the greatest North American mountain. Yet Browne was standing less than 6,000 feet above the sea and 20 miles away from McKinley!

Again the vast wilderness about Mt. McKinley sank back into its natural state. Moose and caribou roamed the foothills, troubled only by wolves, and in the higher crags mountain sheep grazed unmolested. The vast bulk of The Great One towered white and enormous into the sky, streaming with sun and cloud.

Yet this was no world of silence. The millions of tons

of ice which poured from the mountain's shoulders boomed and cracked like artillery in the deep glacial beds. From the sheer cliffs whole mountains of snow shuddered downwards in thundering avalanches, with ice-dust rolling before them in clouds thousands of feet high. Far above, the smooth crest of McKinley gleamed tranquilly in the sun as if to say, "Come try me if you wish, but look below and see what you must contend with."

Two years after their second failure, Professor Parker and Belmore Browne, together with Arthur Aten and Merl La Voy, were ready for another try. This time they started in January, when there could be no travel up the rivers by boat. The glacier waters of the Susitna and Chulitna were buried in silence under many feet of ice and snow. The expedition traveled by dog team up the gorges which they had once ascended in a motor launch.

On the 19th of February, 1912, they left the tiny settlement of Susitna for the journey of more than 100 miles to the northward. On the way they hoped to find a pass which would lead them across the great range to the northeast side of Mt. McKinley. The men in the expedition believed that only from that direction could the mountain be climbed.

Almost a month later, on the 13th of March, they found what they were looking for—an unnamed and unknown river winding down through a deep canyon from the northwest. There was no doubt that this river would take them in the direction of the mountain's northeast face, but no one could tell what obstacles would be met on the way.

The men were still a good 40 miles from the place where they hoped to find the pass through which they could begin their climb. If, instead, they found the sheer walls of rock and ice which had stopped them on their two earlier expeditions, they would again be defeated.

Now you should remember that mountain climbing in the American North is not like mountain climbing in the Mt. Everest region north of India, where native porters may be hired to carry food and equipment. In Alaska the climbers had to be their own porters once they reached a point beyond which dog teams could not go.

On March 13, when the Parker-Browne Expedition reached the unnamed river leading toward the Alaskan Range, it had supplies and equipment sufficient for four men for four months. All of this had to be freighted over glaciers and jagged snow slopes, in zero weather, against stinging, blinding snow for more than 40 miles, with neither map nor trail to guide men and dogs. In some places, the going was so steep that the sleds had to be lowered on ropes fastened to ice-axes driven into the snow. After three weeks of heart-breaking advance under impossible weather conditions, the party found

38

In some places the going was very steep

and crossed through the highest point of a pass. This was a gap 6,000 feet high in the barrier of the great range.

One week later, descending now, they saw below them something they could identify—the great Muldrow Glacier, named after the topographer who first described it. This huge river of ice flows in a stately curve from the shoulders of the great mountain to the low forested land on the northern edge of the Alaskan Range.

By April 24, they had skirted the Muldrow Glacier and crossed the streams which issue from it. Then again they began to climb, pulling their 600 pounds of freight to a height of 2,500 feet. Here above the tree line, among stunted alders and cottonwoods, they made their base camp, with the peak of Mt. McKinley only 20 miles away! Now they rested up for the final attack, eating the fresh meat of mountain sheep that Browne was able to kill. Occasionally they made trips toward the mountain to study the best route to the top.

On the 28th of April they were ready. Their route for nearly ten miles would be over the dangerously tumbled and crevassed ice of the upper Muldrow Glacier. The supplies and equipment could not have been carried that far in sufficient time without the help of dogs.

At nearly 10,000 feet altitude they camped in a bitter snowstorm on a level spot just below a gigantic ice-fall close to the head of the glacier. Here they were forced by the weather to wait. It was in this camp that the first warning came. The mountain was calling up all its forces to keep itself unsullied by man. Let Belmore Browne tell it in his own words:

It was after lunch; Professor Parker was sleeping and La Voy and I were talking in whispers while we listened to the rattle of storm-driven snow across the sides of our frail shelter. Suddenly we felt the glacier under us give a sickening heave and the near-by mountain thundered with avalanches. For an instant I thought that an ice cave had broken in with us, or that the serac [a pinnacle in the ice-fall above them. Ed.] was falling and taking us with it! But in a moment we were undeceived for another shock came, and as the thought of earthquake flashed through my mind the air thundered and pulsated under the force of the countless avalanches. It was an awful and terrifying sound, and we were glad when the echoes ceased and we heard once more the dreary sound of wind and snow.

It was the 5th of June before everything was ready for the final attack on the summit. Supplies had been relayed to 11,000 feet. Beyond that everything necessary for the higher climbing had to be carried on the mountaineers' backs. The weather, however, failed to cooperate, late in the season as it was.

On June 8, after three days of paralyzing snowstorm, the party was terrified by extraordinary noises that seemed to come from the glacier on which they were

41

standing. The ice appeared to groan and crack and echo with a heavy booming sound like the blast of great guns at a distance. If the men had known the cause of this sound, they would have had real cause for worry. Even when they found unexplained ashes in the pot that was used to melt snow water for tea, they did not understand.

Although the noises seemed to come from the glacier, their cause was 300 miles away, on the Alaskan Peninsula. There, in the southwestern part of the territory, an ancient volcano named Katmai, some 7,500 feet high, suddenly blew up on June 8, 1912, in one of the greatest eruptions of modern times. This was surprising, for the volcano had not been active for many years. It was the shock of this catastrophe which shook the rocks and glaciers of Mt. McKinley, causing dangerous avalanches. The eruption may also have had something to do with the fiercely unseasonable weather which was the cause of the climbers' troubles and of their ultimate defeat.

In spite of continued snow and cloud and constant danger from avalanches Parker, Browne, and La Voy reached the great northeastern ridge on June 19, and made a trial climb. (Aten had returned to the base camp, with the dog teams.) This trial ascent carried them to 13,200 feet along a knife-edge of snow which dipped on either side at an angle of 50 to 60 degrees for 2,000 feet on one side and 5,000 feet on the other.

The trial ascent carried them along a knife-edge of snow

The ridge was so steep that Browne had to chop off its crest to make a level place on which they could stand.

At first they believed that it would take less than a day to relay their supplies up into the hollow, known as the Great Basin, which separates the north and south peaks of the mountain. It was actually three days before they reached the entrance to the Basin. There, in addition to the ordeal of further blizzards, they were tortured by their discovery that at 15,000 feet the chief article of their diet—pemmican, a mixture of lean and fat meat—no longer agreed with them. This was serious, for they had no other food to supply the necessary energy and protection from the cold, which at 16,000 feet at 7:30 P.M. on the 26th of June was 19° below zero.

On June 27, in spite of the improper food and an inability to sleep restfully in the awful cold, they believed that the next day would see them on the summit of the mountain. The worst of the climbing was over. From the Great Basin the upward slope was almost gradual. It looked easy.

At 19,000 feet the last bare rock was left behind, and for the first time they could see the summit above them. Yet it was necessary to chop out steps in the icy snow. Close to 20,000 feet, where it was so cold that their hands and feet were almost senseless, the three men felt the icy blast of another blizzard coming upon them. Through the stinging snow they could see that the slope

ahead was much less steep. It was the beginning of
the actual summit! The wind kept increasing, howling
steadily in their frozen faces.

Browne, who had been leading, went back to his companions and the three chopped out a seat in the ice. No
sooner had they dropped into it than they knew they
could not stay there. They were beginning to freeze.
Professor Parker wanted to go on but Browne knew
that it was impossible. He pointed silently to the line
of their ascent. The steps they had cut were already
filled with snow.

Slowly the three groped their way downward through
the storm, feeling for the steps with the tips of their
ice-axes. They reached their highest camp and its comparative shelter between 7:30 and 8:00 that evening,
worn out by their struggle against the freezing wind of
55 miles an hour in a temperature of 15° below zero.

They remained in camp the following day, for the
new snow was not safe for travel. They spent the time
talking about food and weather and trying to dry out
their ice-filled clothing.

At three o'clock in the morning on July 1 in broad
daylight, they set out for the summit once more, but
as they climbed their hearts sank. Rolling up the Susitna
valley from the south was a dense black mass of cloud.
The climbers reached 19,300 feet before the blizzard
darkness once more took charge of the mountain. Denali,

The Great One, had triumphed again! Three heroic men stumbled blindly down through the storm, disconsolate and numb. After four and a half months of struggle, they had to admit defeat.

It was final defeat for Parker and Browne. Yet neither of these pioneers on the great mountain could feel that their defeat was a failure. After all, they had explored and opened up a route to the top and they had reached the top of the mountain. Their only failure was in not having reached the very highest part of the top—the actual summit.

On July 4 the weary climbers reached their base camp where Arthur Aten and the dogs were waiting for them. Here they rested while they sorted out and dried their equipment. On July 6 they were sitting in camp discussing the weather, which had an ominous look, when suddenly a noise like thunder came from the direction of Mt. McKinley. A sudden mist covered the mountains, which seemed to be shrieking and bellowing.

Then the camp collapsed about them. Everything movable, including the stove, was overturned. In front of the tent a huge rock weighing hundreds of pounds broke loose from the earth and came to a stop several feet away. The earth rolled, waved, and shook like jelly. The entire western face of Mt. Brooks, a 12,000-foot peak just east of McKinley, detached itself from the mountain and slid like a gigantic avalanche into the

47

Another blizzard came upon them

glacier valley below it. A few minutes later a cloud of ice-dust and snow rose thousands of feet into the air and rushed down the valley toward the camp of astonished mountaineers.

The men just had time to anchor the edges of their tent with rocks and to crawl beneath it when the cloud, hurtling downwards at 60 miles an hour, struck them and roared on down to the lowlands. After it had passed they crawled out and looked about them in amazement. The streams below them had become chocolate-colored with mud and were flooding their banks. Above them, mountains which had been pure white were now seamed and scarred and half-hidden in a tawny cloud of ice and rock dust.

The four explorers looked at each other in silence. They knew that if they had been on the mountain they would no longer be alive. They had, by a mere accident of timing, lived through one of the world's great earthquakes. The trembling of the earth would have wrecked a modern city and killed most of its inhabitants—had there been such a city within 100 miles.

On their return trip, the climbers heard rumors that only a few weeks before, the mountain's north peak (about 1,000 feet lower than the true summit) had been climbed by prospectors from Fairbanks, 150 miles to the north. Having seen the north peak and learned what climbing conditions were on the mountain, the explorers doubted the rumor.

With the defeat of the best-equipped and hardiest force yet sent to climb it, Mt. McKinley had exhausted its bag of tricks. When Dr. Hudson Stuck, Archdeacon of the Yukon, set out in the spring of 1913, he was not aware that he was approaching a mountain whose days as an unclimbed peak were numbered.

Dr. Stuck was a small, wiry man weighing not much over 150 pounds, but he was an experienced mountain climber. He had spent many years in the Alaskan snow and ice which, as he said, couldn't change much just because it was lifted 20,000 feet in the air.

Two of Dr. Stuck's companions on the actual climb were Alaskans who had packed and sledged over thousands of untracked miles in all parts of the vast territory. The fourth man was a young missionary who had had two years' experience in the North.

These men knew and were able to use the country better than the members of the Parker-Browne expedition. They did not freight in unnecessary quantities of pemmican because they knew that at their base camp they would be able to find game. This could be prepared as food on the spot.

Dr. Stuck's headquarters were at Nenana, 50 miles southwest of Fairbanks and a little more than 100 miles from Mt. McKinley. From this point, the Doctor and his

Dr. Hudson Stuck

party approached the mountain from the north. Early in April, he arrived at his base camp in the foothills, near where Parker and Browne had set up theirs. With him were two Indian boys from the Nenana mission—Johnny Fred and Esaias—who were to drive and care for the dogs. From the base camp Esaias was sent back to Nenana with one of the teams. Johnny Fred remained at the base camp with the other team, and although he was only 15 years old, he also kept house, hunted, and cared for the dogs.

Dr. Stuck's trail up the mountain was almost exactly the same as the Parker-Browne route. Like the earlier expedition, the Doctor used dogs to freight supplies up the Muldrow Glacier as far as was possible. From then on, the men had to carry loads on their own backs, deposit the supplies in a cache, and then return for more. All this was done through snowstorms, bitter cold and—strangely enough—hot, blazing sun. Loads had to be carried over glacier ice that was pierced by snow-hidden crevasses hundreds of feet deep. Dr. Stuck estimated that each one of his group of four climbers actually climbed more than 60,000 feet, instead of the 20,000 feet that was the distance to the summit.

The necessity of carrying things to as great a height as possible, leaving them in a camp, and going back for more, was very nearly the undoing of the Stuck expedition. One day they were returning from below, heavily loaded, to a camp already established higher up when they saw smoke on the glacier above them. The smoke was coming from the site of their camp!

Now in the North, above the tree line, smoke can mean only one thing—the presence of man. But what man could be up there at the camp? Dr. Stuck and his party hurried on and arrived at the cache just in time to save a fraction of their supplies. Their tent was ablaze and all their precious sugar, powdered milk, baking powder, dried fruit, pilot biscuit, most of their tobacco, spare socks, gloves, and camera film had already been destroyed. The smoke had, it is true, indicated man,

but the man was one of themselves who had carelessly thrown away a match without making sure it had gone out.

On May 9 the dogs were sent back to the base camp with Johnny Fred, and the climbers set out to climb the great northeast ridge leading from the Muldrow Glacier to the upper basin between Mt. McKinley's two peaks. They had read a magazine article by Belmore Browne describing this ridge. Browne spoke of it as "a steep but practicable snow slope," and that is how it appeared in the photograph printed with the article. Yet the ridge which lay before them was not at all like the ridge in the photograph! It was a broken and tangled mass of ice and rock that resembled the teeth of a saw.

Dr. Stuck stood aghast, unable to believe his eyes. Then he understood. The earthquake, which the Parker-Browne party had reported—the worst earthquake since the San Francisco disaster of 1906—had made the only known route to McKinley's summit almost impassable.

The Doctor and his companions pitched in and cut a way through the jumble of ice-blocks huge as houses. It took them three weeks to ascend the ridge which Parker and Browne had climbed in two days. That journey of three miles—one mile a week—stands as one of the most remarkable achievements in the history of mountain climbing. It actually broke the back of McKinley's resistance.

53

The smoke was coming from the site of their camp!

Yet the great mountain still had a few strings left for its bow. Having tried practically everything else to discourage its attackers, it now tried heat. The weather had settled down to a steady, almost unbroken brightness with no serious storms, but on June 4 the thermometer rose to 50°, which is seriously warm on the sun-reflecting ice and snow. At that temperature, packing loads in deep snow is heart-breaking, back-breaking work.

Yet even the heat and glare and sometimes the cold—for it still dropped occasionally to 10° and 20° below zero—did not stop the climbers. The weather was unusually clear and they had plenty of food. They crossed the Great Basin in comparative ease and camped at the base of the final ridge with only one day's climb ahead of them and supplies sufficient for three weeks. It was hard to breathe but not impossible. Dr. Stuck, much the oldest of the four, could only climb for a minute or two without resting for several more.

On they went, favored by weather which, though clear, was again bitterly cold. It was so clear that they could see, on the very summit of the North Peak, a wooden flagpole! So, the rumor they had heard was true. The hardy Alaskan prospectors from Fairbanks had been there after all.

The prospectors had climbed from the glacier, it afterwards turned out, in a single day, with neither ropes nor ice-axes to help them, carrying with them a 14-foot

54

flagpole! This they had managed to plant on the summit of the North Peak, which Belmore Browne had thought no man could climb. True, they did not get within 800 or 1,000 feet of the summit of the highest mountain in North America, but there have not been many such climbs in all human history. No one knows why they chose the North Peak. Perhaps they believed they had climbed to the very top of the mountain. Perhaps they had deliberately chosen the North Peak because it, and not the higher South Peak, could be seen from their home town of Fairbanks in clear weather.

Dr. Stuck and his companions were relieved to find that they did not have to climb the North Peak to prove whether or not someone had reached it. They pushed on, foot by foot, up the last pitch of the great mountain. There were moments when Dr. Stuck felt that he could not go on, but he never gave up.

One thing became plain. Although the Browne-Parker expedition of the year before had reached a height of more than 20,000 feet on the actual crown of the mountain, they had been a great deal farther from the true summit than they supposed.

The last long mile, with nothing but snow in sight, seemed endless. Yet the end came and The Great One, though no less great, was no longer unclimbed. The Stuck Expedition had reached the very top of Mt. McKinley!

They rose to admire the view

After a brief prayer of thanksgiving, the weary climbers sprawled in the snow. They made observations of temperature and barometer readings, took bearings with a prismatic compass, and made, as well as freezing fingers would allow, a few photographs. Not until this work was completed did they rise to admire the view.

Dr. Stuck, after the climb, described his feelings of awe in that high place:

Only those who have for long years cherished a great and almost inordinate desire, and have had that desire gratified to the limit of their expectation, can enter into the deep thankfulness and content that filled the heart upon the descent of this mountain. There was no pride of conquest, no trace of that exultation of victory some enjoy upon the first ascent of a lofty peak, no gloating over good fortune that had hoisted us a few hundred feet higher than others who had struggled and been discomfited. Rather was the feeling that a privileged communion with the high places of the earth had been granted; that not only had we been permitted to lift up eager eyes to these summits, secret and solitary since the world began, but to enter boldly upon them, to take place, as it were, domestically in their hitherto sealed chambers, to inhabit them, and to cast our eyes down from them, seeing all things as they spread out from the windows of heaven itself.

When the climbers left Johnny Fred and the dogs on the glacier they had told him they would return in two weeks. It was now going on four. When they finally reached the base camp they had been absent for 31 days.

Fifteen-year-old Johnny had been faithful to his trust. He had kept the dogs well, and the camp was spick-and-span. This Indian boy, whose only schooling had been that of the wilderness and the mission school, brought nobility and honor to the mountain whose name had been somewhat sullied by the college-bred Dr. Cook.

Johnny had been with the four climbers on the glacier when they discovered their tent on fire and their supplies destroyed. He knew that after their ordeal on the mountain they would have neither sugar nor milk. He resolved that they should have both. For 31 lonely days he left untouched the supply of sugar and milk which had been left for him in camp. It was there waiting for the explorers when, worn and tired, they reached their base.

It is hard to believe that Dr. Cook's pleasure in claiming a false title as conqueror of Mt. McKinley was as great as Johnny Fred's at giving up, for the benefit of others, something which was rightfully his.

3. The Mystery of the Missing Pharaoh

The Discovery of the Tomb of Tut-Ankh-Amen

A great deal of time and thought and plain hard work have gone into the building of our communities, with their churches, railways, airports, theaters, libraries, factories, and harbors. They look so solid and enduring that we feel they will last forever. It is impossible for us even to think that some day these landmarks may

59

become buried deep in earth and that men of the future may build a different world above the remains of ours.

Very likely the young people who lived beside the River Nile in Ancient Egypt more than 3,000 years ago felt the same way. They probably believed that the glory of their kings could never fade and be forgotten, that their great temples and palaces could never be lost and buried in the sand.

The Ancient Egyptians were a proud and noble people. Among them were great architects, artists, and engineers. Their powerful kings lived in the midst of a gold and jeweled splendor unlike anything in the world today. Yet there is nothing left of this ancient people. The last of them have long since been absorbed by an alien race of conquerors. Only the greatest and most strongly built of their monuments remain, for the most part half in ruins, for historians to puzzle over and for tourists to wonder at.

The man of modern times, however, has not been content merely to look at the great pyramids near Cairo, which not even thousands of years of blowing sand have been able to wear down or cover, and simply say, "Well, well! I wonder what those great things were for. Who do you suppose would build a thing like that out here in the desert? I suppose we'll never know!"

Slowly, over the years, the story of this great and ancient race has been pieced together. Fragments of its history have been found in old records, in translations of inscriptions carved on stones and ruined buildings, and even on pieces of paper made from water plants. In these documents the Egyptians themselves wrote accounts of the manners and customs of their people and kings.

As the years went on people began to realize that study of the lives and histories of ancient peoples might help us to keep our own civilization from perishing as all others have done. A more thorough study of the past was begun, but the gateway to times that are gone is a dark and difficult one. Listen to the story of one man's exciting adventure into the mystery of Egypt.

George Herbert, Fifth Earl of Carnarvon, had always wanted to dig for relics of the past—had always been interested in archaeology. (The term "archaeology" is taken from a Greek word for "old," and means the knowledge or science of the remains of ancient man and his possessions.) By 1906 Carnarvon had settled upon Egypt as the place in which to satisfy his desire.

Nearly 500 miles up the Great River Nile from Cairo, in a hilly valley of red rocks, golden sand and clear, burning sunlight, lay the half-buried remains of one of the great cities of antiquity. This was Thebes, the capital of Egypt during the time of her greatest power. Long ago the great temples of Karnak and Luxor, cities that stand on the site of ancient Thebes, had been cleared of sand and rubbish for the world to wonder at. Still, at the beginning of the twentieth century, the ancient burial places of Thebes held secrets that had never been disclosed.

During the past 150 years it gradually had become known that the most important Theban burial place lay west of the Nile and some distance from the town of Luxor. It was also known that when an Egyptian king, or Pharaoh, died, all his belongings were buried with him.

Some explorers of the region about Thebes had been fortunate enough to locate ancient burial places. Of the 30 or more graves of Pharaohs thought to lie in the Valley of the Tombs of the Kings, some 28 had been located and dug out. Mummies and great stone coffins, jars, vases, and other objects were found, but without exception every royal tomb was found to have been entered by robbers who had removed the most valuable treasures.

Lord Carnarvon and his archaeologist helper, Howard Carter, knew that the tomb of one king who had lived about 3,300 years ago had not yet been found. Carnarvon decided to concentrate on an attempt to locate the missing Pharaoh. There was a faint hope that since exploring archaeologists had not been able to find his burial place, thieves also might have missed it.

This king was a mysterious figure about whom history

Lord Carnarvon *Howard Carter*

has told us very little. Even his name, which we now know to have been Tut-Ankh-Amen, was not entirely certain when Lord Carnarvon and Howard Carter began looking for him. It was known that he was young, and it was supposed that he was not of the royal line himself, but that he had become a king by marrying the youngest daughter of the great Pharaoh Akh-en-Aten. But where was Tut-Ankh-Amen buried?

Howard Carter had worked some years before with an American archaeologist to whom the Egyptian Government had given the right to dig in the Valley of the Tombs of the Kings. In the course of this work, Carter found a few objects marked with the name of Tut-Ankh-Amen, some of them in a small tomb in a neglected part of the Valley. The American, Mr. Theodore Davis, claimed that he had found the tomb of Tut-Ankh-Amen but that there was no king in it; the tomb had been plundered like all the rest.

Howard Carter knew better. He was sure that the tomb found by Mr. Davis was not a royal tomb. No king of the 18th Dynasty would be buried in such a commonplace grave. Then the discovery of some pottery jars containing materials used during the funeral of Tut-Ankh-Amen made Carter feel that the real burial place of that king could not be far away.

Let us take a good look at the Valley of the Tombs of the Kings and see what a discouraging task Carter had set himself. If you stand at the entrance to the valley you will see before you a great rock wall. Above it rises a rocky peak known as "The Horn." At the foot of the rock cliff, fans of fallen debris, rocks, and dusty earth slope down to the valley floor. In some of the slopes are stoned-up openings where tombs have been found and excavated, the rubbish from them being dumped on the floor below.

Howard Carter realized that it would take years of digging and hundreds of men to move the debris and get down to the bedrock in which the tomb, if it were there at all, must have been built. Was there no clue which

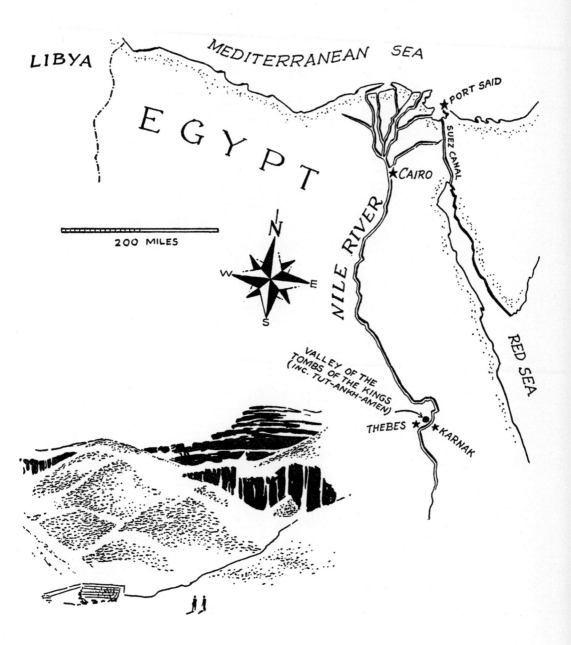

LIBYA

MEDITERRANEAN SEA

EGYPT

★ PORT SAID

SUEZ CANAL

★ CAIRO

NILE RIVER

200 MILES

N
W E
S

RED SEA

VALLEY OF THE
TOMBS OF THE KINGS
(INC. TUT-ANKH-AMEN)

THEBES ★ ● ★ KARNAK

might show one spot to be more likely than another? To Carter's trained eye there was.

Below the open entrance to the long-since discovered tomb of Rameses VI, a season's digging unearthed a group of crude huts. In huts such as these, the ancient workmen who built the tombs in the valley had lived thousands of years before. Furthermore, the huts were erected over a great pile of flint boulders.

Now Carter knew from experience that the presence of such boulders often meant that there was a tomb not far away. Unfortunately, to dig in the neighborhood of those boulders would have cut off the path to the tomb of Rameses VI, and that tomb was very popular with visitors to Egypt. So Carter decided to wait until the tourist season had passed, and work on the site was suspended until October, 1922.

By November 3 a number of the ancient workmen's huts had been removed. They were all more or less alike and enough were spared to preserve their historical relationship to the tomb of Rameses. Beneath the spot where the huts had stood were three feet of soil and below that, bedrock. The difficult work of clearing away this dry and dusty earth began.

When Howard Carter arrived at the site on November 4, he looked about him anxiously. There was something wrong. Instead of the usual clatter, scrape, and clang of picks, hoes, and shovels and the chatter of workmen's voices, the Egyptian sun fell upon a scene of absolute silence. Something had happened. What was it? Good or bad? Carter was used to the bad. It was what he expected.

Ready for the worst, he studied the face of the foreman who came forward to speak to him. What he heard seemed too good to be true. That very morning, he learned, the workmen digging beneath the site of the first ancient hut to be destroyed had come upon what appeared to be a step cut in the living rock of the valley floor. This could not be modern work, since the huts which had just been removed from above it had been standing there since the death of Rameses VI, more than 3,000 years before. Carter dared to hope that he had really found something after all his years of searching.

For two days his men worked to clear the spot, following the lines of the step until all four sides of what was plainly a stairway appeared in the rubble. There could now be no doubt that what had been unearthed was the entrance to a tomb! Yes, but how many times had eager diggers slaved to clear tombs, only to find that they were unfinished or had never been used!

The digging went on, until 16 descending steps in a passage ten feet high and six feet wide had been cleared. Howard Carter held his breath as he watched the passage being opened. What he said when he saw the upper part of a doorway sealed tight with plaster and stone has

A stairway appeared in the rubble

not been recorded. Here at last, after years of dis-
appointment, was a real tomb. A tomb, yes, but whose
was it? There was one way to find out.

When most Theban burial places were closed up, they
were marked with two seals. One was the seal found on
all tombs in the royal cemetery, indicating the presence
of a very important person. The other was the personal
seal of the Pharaoh whose body lay within. Carter

searched the unearthed doors for a royal name. He found only the first type of seal.

While studying the ancient plaster, however, he saw at the top of the door, where some of the plaster had fallen away, a piece of heavy timber. It was apparently the lintel of the door. Here was a chance to see what lay beyond.

Carter's heart was beating violently as he poked at the plaster below the wooden lintel, making a small hole through which he could point the beam of his flashlight. He hardly dared to follow the light with his eyes, but inside he saw nothing—nothing but stones and rubble. Yet that very nothing might mean everything! It could mean that the utmost care had been used to make the tomb difficult to enter. If there was nothing hidden beyond, no one would have bothered to fill up the passageway from floor to ceiling with stones and then seal the door which led into it.

Carter now knew that he had possibly come upon the most remarkable discovery ever made in Egypt. It was plain that this was an important burial place, plain too that it had not been molested by thieves for at least 3,000 years. But what of the tomb itself? It was, compared with the tombs of Pharaohs already discovered in the valley, a small and insignificant opening. Perhaps it was not a royal tomb at all but that of some nobleman buried in the valley by a king's consent. Carter turned to

69

the door again and once more searched it for identifying seals. As far down as it had been excavated he found none.

What was he to do? It was growing dark. The work of clearing the door could not be finished that night. Neither could he leave exposed what had already been brought to light. Reluctantly, with no one but native workmen about him, he filled in the small peephole which he had made below the top of the door frame. Then he gave orders to have everything put back in place.

After he chose the most trustworthy of his men to stand guard for the night, Carter went home through the eerie moonlight to sleep. That was no easy task. There was too much to think about, for nothing more could be done until the arrival of Lord Carnarvon from England. After all, Carnarvon was the sponsor of the expedition, and it was his right to be present at the opening of the tomb.

Carter tossed and rolled, got up and walked the floor of his room. He knew that he must wait. In the morning he sent a cable message to Carnarvon. Then he went back to the valley and devoted himself to the heartbreaking job of filling in and smoothing over the entire site of the new and exciting excavation. By evening of November 6 it would have been impossible for a visitor to tell that any digging had ever been done there. But the news was out. Carter began to receive telegrams of congratu-

70

lation, letters casting doubt upon the value of the discovery, and offers of help from all over the world.

During the more than two weeks which passed before Lord Carnarvon was able to reach Egypt, Howard Carter enlisted a staff of skilled assistants to help with the work of clearing the tomb. The tomb itself remained as it had been for 30 centuries, its mystery still unsolved, like a facedown pack of cards from which might possibly be drawn an ace or, as Carter knew only too well, a joker.

On November 23, the work of re-opening the tomb entrance began, for Lord Carnarvon and his daughter, Lady Evelyn Herbert, had arrived in Luxor across the Nile from the Royal Valley. By the following afternoon the doorway was again exposed, this time all the way to the bottom. Once more a search was made for signature seals, but this time the name of Tut-Ankh-Amen could be made out in several places. Carter for the first time allowed himself to hope that he had found what he was looking for.

However, some unpleasant possibilities still remained. In the first place, a study of the sealed door showed that an opening, large enough to admit a man's body, had been made at some earlier date—probably within 100 years of the burial—and then sealed up again. Yet the very fact that it had been sealed probably meant that some valuable articles remained within the tomb.

There was one other puzzling and discouraging fact. In the rubbish found on the stairway that led to the door were fragments of pottery, boxes, and other objects marked with the names of half a dozen kings. This might possibly mean, thought Carter, that what he had found was a storehouse rather than a tomb, a storehouse in which the possessions of many Pharaohs had been put for safekeeping, probably during Tut-Ankh-Amen's reign. If this were true, then of course Tut-Ankh-Amen's body would not be found behind the well-sealed door.

By November 25, Carter's hopes had all but disappeared. During the morning of that day the seals were photographed, and then at last the blocked door was cleared away. Now the diggers were able to see a descending passageway without steps. As Carter had already noted through his peep-hole, the passage was packed with stone and rubble from floor to ceiling.

The members of the expedition saw something else. There were signs that Carter was not the first to penetrate the mysterious underground darkness which lay ahead. Someone, in ancient times, had burrowed through the debris just under the ceiling in the upper left side of the passage. On coming out again, this person had filled up his burrow and tried unsuccessfully to make it look as if no one had been there. So there it was, like it or not. The spot had been plundered, like all the others. What, then, could be expected?

72

All that day and until mid-afternoon of the next, Carter watched anxiously as he helped his workmen clear the passageway. It was slow going, for every basketful and barrowload of material carried out had to be examined for telltale objects. At last the outlines of another door, exactly like the outer one, appeared. Once again, Carter felt the thrill of excitement. That door, when opened, might bring to view—for the first time in modern history—the details of the most fascinating and least known period of Egypt's greatness.

It was hard for Howard Carter to believe, after so many disappointments, that he might really be standing before the gateway to another world. It seemed to him that the workmen were unusually slow in clearing the second door. When at last the whole height of it stood unobstructed before him the next step seemed almost too great a one to take. Suppose that after all this preparation there should be nothing? He looked at Lord Carnarvon, whose expression did not change.

Then Carter approached the door. His hands trembled as with hammer and chisel he made a small hole in the door's upper left-hand corner. The sound of the clinking blows echoed in the underground chamber in which no sound had been heard for 3,000 years.

When the hole had been drilled all the way through, Carter lifted a long iron bar and poked it slowly as far as he could reach into the darkness. It touched nothing, showing that at least the chamber beyond had not been filled up with earth and stone as had the passages which led to it. Hot stale air, thirty centuries old, issued gustily from the hole. A candle flame held in the draught flickered and sputtered but did not go out, indicating that the air within the closed room was at least breathable.

Carefully Carter enlarged the hole until he could get his arm, still holding the candle, through it. The candle flame, still flickering, kept him from seeing as he bent his head and peered along his arm. Then, as his eyes became adjusted to the light, he saw—at first dimly, then more clearly—what surely no man of modern times had ever seen. Was it the candle's glitter that made everything seem to be of shining gold? No! Gold it was.

Gold was everywhere. Curious beasts of gold! Gilded statues and boxes! Pieces of furniture of strange and beautiful shape, shining with gold and soft with the paler hues of ivory and alabaster!

For a moment Carter was speechless, motionless. The others in the passageway behind him held their breath. Carter stared. Then he heard Lord Carnarvon behind him whispering, "Can you see anything?"

Carter withdrew his hand from the hole, passed it before his eyes. For a moment he could not speak. Then he said, well aware that the words were almost meaningless, "Yes, wonderful things!"

The candle flame did not go out

When the hole in the door had been chipped out so that two people could look through, Carnarvon and Carter, side by side, marveled at the form and beauty of the treasure they had found. Let Carter describe the scene in his own words:

Gradually the scene grew clearer, and we could pick out in-dividual objects. First, right opposite to us—we had been con-scious of them all the while, but refused to believe in them—were three great gilt couches, their sides carved in the form of monstrous animals, curiously attenuated in body, as they had to be to serve their purpose, but with heads of startling realism. Uncanny beasts enough to look upon at any time: seen as we saw them, their brilliant gilded surfaces picked out of the darkness by our electric torch, as though by limelight, their heads throwing grotesque distorted shadows on the wall be-hind them, they were almost terrifying. Next, on the right, two statues caught and held our attention: two life-sized fig-ures of a king in black, facing each other like sentinels, gold kilted, gold sandalled, armed with mace and staff, the protec-tive sacred cobra upon their foreheads.

These were the dominant objects that caught the eye at first. Between them, around them, piled on top of them, there were countless others—exquisitely painted and inlaid caskets; ala-baster vases, some beautifully carved in openwork designs; strange black shrines, from the open door of one a great gilt snake peeping out; bouquets of flowers or leaves; beds; chairs beautifully carved; a golden inlaid throne; a heap of curious white oviform boxes; staves of all shapes and designs; be-neath our eyes, on the very threshold of the chamber, a beauti-ful lotiform cup of translucent alabaster; on the left a confused pile of overturned chariots, glistening with gold and inlay; and peeping from behind them another portrait of a king.

Such were some of the objects that lay before us. Whether we noted them all at the time I cannot say for certain, as our minds were in much too excited and confused a state to register accurately. Presently it dawned upon our bewildered brains that in all this medley of objects before us there was no coffin or trace of mummy, and the much-debated question of tomb or cache began to intrigue us afresh. With this question in view we re-examined the scene before us, and noticed for the first time that between the two black sentinel statues on the right there was another sealed doorway. The explanation gradually dawned upon us. We were but on the threshold of our discovery. What we saw was merely an antechamber. Behind the guarded door there were to be other chambers, possibly a succession of them, and in one of them, beyond any shadow of doubt, we should find the Pharaoh lying.

We had seen enough, and our brains began to reel at the thought of the task in front of us. We reclosed the hole, locked the wooden grille that had been placed in the first doorway, left our native staff on guard, mounted our donkeys and rode home down The Valley, strangely silent and subdued.

The following day, November 27, was largely taken up with the clearing of the door to the treasure chamber and making a hurried study of the magnificent things it contained, none of which could be touched or moved until they were photographed, numbered, and listed. This study showed that most of the articles in the room were marked with the name of Tut-Ankh-Amen. Now Lord Carnarvon and Carter had added reason to be certain that what they had uncovered was indeed the tomb of the missing Pharaoh.

Archaeology is a hard taskmaster. Sticking strictly to its rules, an explorer of old ruins cannot gallop from

one thing to another, following what interests him most.

Howard Carter would have liked to tackle that last mysterious door without delay but he knew that he must not. First he must take care of the amazing objects gathered in the antechamber and in the little annex opening from it. The robbers of 3,000 years ago had made a frightful mess of the annex, without apparently stealing anything of value. Proper attention to the priceless objects required the setting up of a laboratory in a nearby empty tomb. Then a staff of experts would be needed to identify, preserve, and repair the delicate and beautiful things that were being exposed to light and air after thousands of years in the dry, dead darkness of the tomb.

Being a true scientist, Carter had the courage and patience to do what he must do and let the thrill of new discovery wait until the proper time. It was not until February that the great moment came. Then the antechamber and annex had been completely cleared and every ounce of dust from the floor sifted for stray beads, pieces of inlay, jewels, and crumbs of gold. Only two objects had been left—the two guardian statues of the King, one on either side of the still unopened doorway.

On the afternoon of February 16, 1923, a small gathering of very important persons met, by invitation, above the entrance to the tomb. Solemnly they descended to the cleared antechamber, now filled with chairs. At the

79

They marveled at the treasure they had found

north end of the room a platform had been built so that
Carter and his helpers, Mr. Mace and Mr. Callender,
could reach the top of the sealed door. At quarter past
two Carter, with trembling hands, raised his hammer
and chisel and opened what everyone imagined would be
the final scene in the Mystery of the Missing Pharaoh.

Ten minutes later a hole had been made, and the beam
of the electric flashlight showed, not three feet from
the stoned-up door, what looked like a solid wall of
gleaming gold. Excitement seemed to fill the room as
stone after stone was loosened and passed back from
hand to hand and out of the chamber. Lower and lower
went the stones which blocked the door. Higher and
higher seemed to loom the shining gold wall which rose
from the floor of the inner room some four feet below
the threshold of the door.

There was no longer any question. This was certainly
the shrine in which had been buried the remains of the
lost boy ruler of that ancient kingdom. What was still
not certain was whether or not the Pharaoh himself
still lay within it. For plunderers had been there, though
they had apparently left in haste.

On the threshold of the door Carter found the scattered beads of a necklace which the robbers had dropped
as they hurried out. The necessity of stopping work to
pick up these precious beads added to the excitement of
the twenty watchers, who sat squirming on their chairs

before what was probably the richest discovery ever made in Egypt.

It was five o'clock when everything was cleared away from the doorframe. Carter lowered himself into the room in which the shrine stood, and Lord Carnarvon and the Director of Egyptian Antiquities, Mr. Lacau, followed. They soon found that the doors of the great shrine were bolted but not sealed. It was still impossible to tell what damage the ancient thieves had done. Slowly and cautiously the breathless men, too humbly reverent to be classed with the ancient plunderers of tombs, drew the bolts and swung back the doors. Heavy as they were, they moved as easily as if they had been closed only the day before.

Here at last was certainty. Their dream had come true. Within the first set of doors was another shrine. This, too, had bolted doors, but upon them was an unbroken seal proving, without question, that the inner shrine had not been entered since the King was laid to rest.

The three men exchanged glances but did not speak. Beyond those golden doors, they now knew, lay the body of a young man who had once held power over the great empire of Egypt.

Tut-Ankh-Amen's subjects had buried him there, hoping, by means of the sacred charms which they had placed in his tomb, to make his journey to the other

world safe and undisturbed. For this journey they had buried with him all his most intimate belongings, no matter how great their value. Here in the outer chamber had been found the Pharaoh's linen, his sandals, his jewels, his gloves, walking sticks, bows and arrows, his couches and his chairs, even his priceless golden throne. Here was the Egypt of 3,300 years ago, alive and unruined.

The three men moved away from the golden doors knowing what must be done before the King himself could be disturbed. For the sake of future generations, every detail of the incredible collection must be examined and studied. Howard Carter knew that this would be a long and exacting task. What he did not know was that his friend and sponsor would never look upon the face of the young King whom he had helped to rescue from the silent darkness of the past. Less than two months later Lord Carnarvon was dead.

Had Howard Carter known the difficulties ahead, he might have doubted that Tut-Ankh-Amen would ever be seen by modern man. Lord Carnarvon's death and a misunderstanding with the press and the Egyptian authorities, together with the work of preparing the burial chamber, caused a delay of two years before the sarcophagus—or stone coffin—could be opened. In October, 1925, the work of digging out the tomb, which had been filled in so that no one could get at it, began all over again.

Their dream had come true!

During the winter of 1923-24 the golden shrines had been taken apart and moved out of the tomb. The great carved sarcophagus, containing the nested coffins in which it was believed the King's body still lay, was massive and forbidding. It stood in the center of the chamber now stripped of all its glitter and gold.

In the cramped space of the burial chamber, a feat of engineering was required to lift and take away the sarcophagus' lid, which weighed well over a ton. Here at last was the end of the search. But no! Not quite the end. Inside the great stone box the searchers found not the mummy of the King, but a beautiful coffin of sheet gold, heavily ornamented with images of gods and symbolic designs. At the head of this splendid burial case was a mask of inlaid gold, made in the likeness of the dead monarch.

The forehead of the mask was decorated with the emblems of Upper Egypt (the Cobra) and Lower Egypt (the Vulture). Over them had been set, probably by the delicate young hand of the widowed Queen, a small wreath of flowers which still, after 33 centuries, retained their color and a little of their scent. It was hard for those who saw this touching sight to realize that more than 3,000 years had passed since the mourners at Tut-Ankh-Amen's funeral had tiptoed out of the tomb into the blazing sunlight of the Valley.

It was not until November, 1925, that the actual opening of this magnificent coffin took place in the presence

84

Mummy of Tut-Ankh-Amen, boy king of Egypt

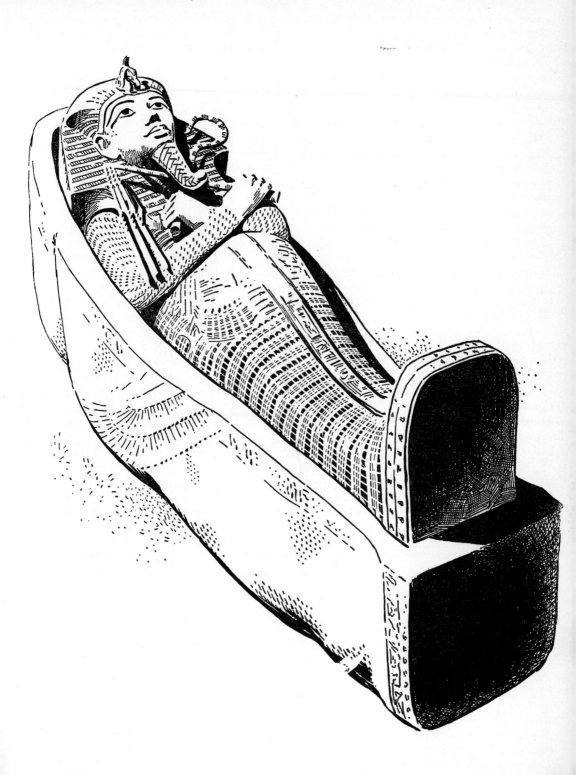

of a few chosen scientists and government dignitaries. The work of clearing the burial place had taken a long time, but everyone involved felt rewarded by what took place that November day. There was no longer any doubt that the body of the Pharaoh lay beneath that wealth of gold. Yet never before had a great King of Egypt been found untouched just as he had been laid to rest.

Actually, there were three body-shaped coffins nested one within the other. The third and innermost, which contained the mummy, was of solid gold almost a quarter of an inch in thickness. The two outer coffins were of wood covered with sheet gold and ornamented with inlay of glass and lapis lazuli.

As the discoverers of this amazing treasure looked on the face of the boy King, they saw that it was still recognizable. The sight must have made many of the visitors feel the ties that bind all human beings—those of the past and those of the present. Through the gateway of Tut-Ankh-Amen's tomb we pass not into history but into the very heart of the life of a boy who lived and breathed and hoped and feared very much as we do today. What, in that tomb, was more important to the discoverers? Was it the treasure of gold and the telltale objects of daily life? Or was it the wreath of flowers on the King's forehead, so real that it seemed to be still warm from the light hand and still salty with the tears of a beautiful young Queen?

4. The Dragon That Did Not Die

The Burden Expedition to Komodo Island

One of the most fascinating things about human knowledge is that it is never complete. At certain times in our lives most of us are inclined to think that for us, at least, there is nothing more to learn. This is somewhat like feeling, after a heavy meal, that we cannot eat another thing. Give us only a few hours! Then once again we shall be hungry.

If you were to believe that every existing form of life on earth has been discovered and named you would not be very likely to organize an expedition for the purpose of hunting dragons. You would find it much easier to go to the natural history books and encyclopedias and discover that there are no such fantastic beasts. That would be your discovery if you were to consult the older of the natural history books published since the beginning of the seventeenth century.

It is true that a celebrated naturalist of the sixteenth century, Konrad von Gesner, who died in 1565, devoted a whole chapter to dragons in his *History of Animals*.

If you had told Dr. Gesner that you did not believe dragons should be considered in a serious scientific book, he would have looked at you, no doubt, in some astonishment. Had not Dr. Cardanus of Pavia, in Italy, actually seen dried dragons in Paris? He certainly had. They were presumably young dragons, for Dr. Cardanus mentions that they were very small.

A Frenchman who lived at about the same time as Dr. Cardanus and Dr. Gesner actually printed a picture of this interesting creature. The animal shown in the picture is not so mysterious or mythical as it might seem. It is in fact a fairly recognizable sketch of a beast which is included in natural history books today under the name—or at least the Latin equivalent of its name—dragon! *Draco volans*, the flying dragon.

88

Now the sixteenth-century naturalists who wrote about dragons were not at all troubled by the fact that the creature they wrote about was assumed to be an enormous, man-eating, fire-breathing partner of giants. What difference did it make if the creature they saw in the dried Paris specimens was less than ten inches long including tail? After all, a dragon had to be hatched from an egg, and you could not get a very large dragon into an egg. These Paris specimens were, they decided, simply baby dragons. The fact that no one had ever produced a full-grown specimen, alive or dried, seemed perfectly natural. The grown dragon, these early naturalists concluded, was smart enough to keep out of man's way!

Scientists, as they came to know more about the classification of animal life, dismissed the giant dragon as a myth and left him out of their books. *Draco volans* they kept, because they knew what he was and where he lived. He was (and is) a small lizard living in the Malay Peninsula and on some of the islands of the Malay Archipelago—Java, Sumatra, and Borneo.

In fact, there are even more dragonlike and far larger creatures living in that part of Australia which is nearest to the east end of the Malay Archipelago. A picture of one of them, which is known to reach a length of three feet, is shown on page 103. If anything could look more like a dragon than that it would have to be a dragon.

It was all very well for scientists to give the mythical

A dragonlike creature found in Australia

idea of a dragon the cold shoulder. Millions of people in many parts of the world believed that they knew better. After all, had the scientists been everywhere and seen everything?

A very clear representation of a beast which could only be described as a dragon was found on a gate in Babylon in company with the image of another beast which was known to be a real one and not a myth. The Chinese used the image of a very convincing looking

90

dragon as the emblem of imperial power. It appeared on the Chinese flag until comparatively recently. There must be, said those who did not have to be as cautious as the scientists, some reason for man's belief in such a creature.

Not until the beginning of the nineteenth century did the science of paleontology begin to take shape. (Paleontology is the study of living things which existed in former geological periods and whose remains have been preserved in the form of fossils.) Before the nineteenth century the discovery of large bones which could belong to no known animal was often taken as proof of the existence of dragons. The Chinese even went to the length of quarrying fossil bones, of which China has a great store, grinding them up, and selling them as powdered dragon bones and teeth. This medicine was long believed capable of curing practically every ailment.

The fact that these bones can now be identified and assigned to animals of the past which are almost as well known to scientists as living animals does not, it is true, prove that no such thing as a full-sized dragon ever existed. There are difficulties, however. The most dragon-like of all, the giant saurians, were apparently extinct many millions of years before man appeared on earth and man would not be likely, therefore, to have any memory of them.

However we know that several times within the last

twenty years a five-foot fish weighing over 100 pounds
and believed to have been extinct millions of years before
man has made its way into fish-nets off the east coast
of Africa. So why not a dragon?

It is of course easy to say that dragons did not live in
the sea but on land and the land has been much more
thoroughly explored than the sea, in spite of Barton's
bathysphere. There are not many places on earth in
which a dragon big enough to be anything but a dis-
appointment could hide.

In spite of this, the idea persisted that ancient animals
supposedly extinct might be hiding in one or more of the
little-known parts of the world. There were still unex-
plored places in Africa, Australia, New Guinea, and even
in the gigantic table lands of the Venezuela-Guiana
border in South America. Sir Arthur Conan Doyle used
this last region as the setting for his *Lost World*, a fascinating story which helped to keep alive the idea that
dragons or worse might still be roaming the earth.

The Lost World was fiction, not fact, and very fantastic
fiction at that. The idea of the survival of prehistoric
creatures, however, was not entirely dead, even in the
minds of scientists.

In 1912, that idea got what might have been a considerable boost. In that year a party of pearl fishermen
anchored in a small harbor of an almost entirely unknown island in the Malay Archipelago. The island was

a bit of volcanic rock little more than 20 miles long and half as wide situated in what is known as the Lesser Sunda Islands, 100 miles south of Celebes and some 600 miles northwest of Australia.

The island was, at the time, so little known that even
the sailing directions and charts were wrong about its
coast. It had, until only a short time before, been entirely uninhabited. There had, however, been rumors sug-
gesting that prehistoric beasts survived there. Whether
the pearl fishermen who visited it in 1912 saw one of
these incredible creatures or merely talked with Malayan
residents who had seen them is not clear. At any rate,
the fishermen carried reports of dragons to Java.

These reports were so convincing that Mr. P. A.
Ouwens of the Buitenzorg Zoological Museum in Java
sent collectors to find specimens of the mysterious monster. The hunters were evidently successful, for Ouwens,
just before the First World War, published the first de-
scription of the mysterious strangers of Komodo Island.
Ouwens' account, however, did not get much attention
from the general public and the War soon caused the
interesting find to be forgotten.

It was not until 1926 that it occurred to anyone to try
to add to the scanty information which Ouwens had provided. In that year a young scientist on the staff of the
American Museum of Natural History organized an expedition to Komodo Island for the purpose of bringing
back specimens of the prehistoric monster to be exhibited
in the Museum. If possible, live specimens were to be
captured for the Bronx Zoo.

The expedition consisted of W. Douglas Burden, his

wife, an experienced hunter named DeFosse who had spent
his life in the jungles of Indo-China, and an expert on
reptiles and amphibians, Dr. E. R. Dunn of Smith Col-
lege. In addition there were a Chinese motion-picture
photographer and a Chinese man-of-all-work named Chu.
In Java the Dutch Colonial Government gave permission
for the expedition to visit Komodo. It also lent Mr. Burden a 300-ton steamer complete with crew and supplies
for the long voyage of nearly 1,000 miles from Batavia
to the strait which separates Sumbawa Island from
Flores.

After a battle with a 13-mile-an-hour current in the
strait, the expedition arrived at an anchorage on the
eastern shore of Komodo early one morning in mid-June.
They found an island that sloped steeply upwards toward
a central mass of volcanic peaks. On the shoulders of
the peaks were deep-green patches of thick and thorny
jungle. The weather was clear and cool, and the explorers
were delighted with what seemed to be a naturalist's
paradise—a spot that was both beautiful and interesting.

Yet Komodo, in spite of its location, is a semi-arid island,
grassy in spots but with very few trees at sea level. Its
lower parts are of sharp volcanic rock and loose boulders.
To find a suitable camping place it was necessary to
climb into the higher land of the interior.

On the morning after their arrival, Mrs. Burden
rested aboard ship, and the Chinese cameraman tried to

recover from his seasickness. Eager to be active, Dr. Dunn and Mr. DeFosse struck out to the north, while Douglas Burden climbed westward over the rocky slopes under a burning sun.

The heat reminded Burden, after the fresh, cool morning, that he really was in the tropics. With torn fingers and frayed shoes he finally reached a stretch of lovely rolling country dotted with palms and groves of bamboo. Among the rocks of this enchanted realm he made an exciting discovery—the unmistakable print marks of a gigantic foot, very much like some of the fossil dinosaur tracks he had seen at home.

When DeFosse and Dr. Dunn reported similar tracks from the northern part of the island it seemed that the *Lost World* had been found at last. And a wonderful world it was, crammed with an astonishing variety of strangely mixed wild life. In the jungle thickets many-colored birds, some recognizable and others unknown to the explorers, twittered, sang, and screeched. Deer leaped through the bamboos, wild boar trotted in the underbrush, yellow-crested cockatoos scolded from the higher trees, and pigeons of all colors flew about, not to mention the presence of a greater variety of poisonous snakes than could be found anywhere else in the world.

While Burden was scouting about the pool near which he planned to establish the expedition's base camp, he discovered a plainly marked trail that was deeply rutted

with the marks of an enormous cloven hoof. He knew
that this could mean only one thing. The tracks had been
made by one of the most dangerous animals in the world
—the great Indian buffalo or carabao, a longer-horned
relative of the vicious Cape buffalo of Africa. This
rather uncomfortable thought made the idea of sitting
in the jungle and waiting for the dragon to appear something more than merely exciting. Meanwhile, there was
other work to be done.

As Burden, still in search of a camp site, pushed his
way through the jungle's leafy barrier toward a second
pool in the lava rock, he was startled to see two ducks
whir upward from the water. Before the sudden noise
of their flight had quieted he heard behind him a tre-
mendous crashing which sounded, he says, as if the entire forest of heavy bamboo were being broken to splin-
ters. As he jumped back into a clearing, he saw a huge
bull buffalo rushing straight toward him at full speed.
Its nose was in the air, its nostrils wide, its great horns
laid back along its flanks like the smoke of an express
train.

Burden wasted little time in making up his mind what
to do. As he had no steel bullets to pierce the tough
hide, he jumped quickly into the jungle and scrambled
up a steep rock. Surprisingly enough complete silence
followed. The buffalo must have stopped. But where?
Burden held his breath. There was not a sound but the

screaming of cockatoos. Then, as suddenly as it had ceased, the thundering of the great beast began again. Slowly, this time, it crashed off into the jungle. Although it had probably never before seen a man, the animal seemed to be satisfied that this one was nothing to worry about.

A few days later Burden completed his organization of the first camp ashore. Then he set off one morning to shoot a deer to feed the native porters he had hired from the Raja of the neighboring island, Sumbawa. After securing the deer he went on uphill toward the region of rocky peaks which he had visited on the first day.

It was about half past nine in the morning when he reached the foot of a slope of broken rock upon which grew tufts of thick, short grass and scattered gubbong palms. Pausing for breath on the slope, Burden heard a scraping sound and a slither of falling rock from above. Looking up suddenly he saw what it was, what he had come 12,000 miles to see.

Dropping to his knees, Burden crept up the slope from rock to rock to a point from which, without being seen, he could, so to speak, look back into the past for 60 million years. There before him was the great head, lolling from side to side. Its yellow, forked dragon-tongue was darting in and out from between fiercely toothed jaws that were more than a foot long. As the ancient and fantastic creature lumbered downhill with the morn-

99

Burden jumped quickly into the jungle

This was what Burden had come 12,000 miles to see

ing sun behind him a great black shadow, ten times his size, preceded him.

From his hiding place Burden studied the beast through his field glasses. The rough, wrinkled skin was black with age and marked with the scars of many battles. It suggested, with its scaly surface, a heavy coat of woven steel. On legs like bent tree trunks the monster kept coming. His nostrils, black holes in an evil-looking head, which only needed fire and smoke to make him a true dragon, dilated; and his piercing black eyes seemed to be searching for something among the rocks.

Burden watched him, fascinated. Then, almost as he watched, the monster did something even more dragon-like. He simply disappeared. Burden crept out from be-

100

hind his rock but he could find no trace of the dragon.

After that, Burden and his companions began setting traps, using wild boar for bait. Stakes were driven into the ground all about the dead boar. Then the stakes were lashed together and the enclosure was covered with foliage, leaving only a large opening at one end. Next, a nearby tree was stripped of its branches and left standing. After a rope was fastened to the top of the tree trunk, it was pulled down by fifteen husky natives. The other end of the rope was then fastened to a trigger in front of the bait and a loop in the rope was spread out in the form of a noose.

A few yards from the trap a shelter of heavy stakes was built and it, too, was camouflaged with branches and leaves. From this point the members of the expedition were to watch for any great beasts that might approach the trap. There was only one hitch. It was apparent from the start that the trap's first visitors would be either females or their young, the huge old males being very wary. While these would be satisfactory as laboratory specimens for anatomical study, the real monsters were the ones the hunters wanted.

To keep small animals from springing the trap and eating the bait, Burden attached to the trigger a rope that was long enough to reach the observation shelter. From there the rope could be pulled, setting off the trap only when the right-sized dragon was at the bait.

Early in the morning the hunters took their places in the blind, or boma, as big game hunters call it, and began their watching and waiting. Luckily the giant lizards behaved as if they were stone deaf, for 12-inch centipedes, whose bite is frightfully poisonous, and scorpions, whose sting is no joke either, kept crawling in to join the watchers. Whenever these unwelcome visitors arrived, the men gave cries of distress and slashed about among the dry leaves of the floor.

The sun had been up for some time when the first of the dragons appeared, a small one perhaps five feet long. He (or she) walked cautiously around the trap, sniffing at the wild boar bait which was by this time somewhat overripe and in just the right condition to interest a dragon. This youngster, however, wandered off without entering the trap.

It was followed by a somewhat larger specimen—a creature that seemed to believe he was dragon enough to carry off the entire wild boar. The bait, however, was securely anchored in the trap. In the midst of his struggle, this middle-sized dragon suddenly stopped and looked up with his wicked jaws drooling. Then he lowered his head and raced off into the jungle.

"Aha!" thought Burden. "That must mean a big one is coming!"

For half an hour nothing happened. Then, as if they were able to detect what the hunters could not be aware

He was staring straight into the face of the monster!

of, the natives who were in the blind began to stir in restless excitement. Burden peered through the back wall of the blind and gave a start. He was staring straight into the face of a monster which, the fossil record shows, had endured almost unchanged for 50 or 60 million years.

The dragon stood perfectly still, his great black eye staring fiercely at the blind. Slowly he began to move his long clawed feet. Heavily, cautiously, he stalked past the blind with its quivering occupants and headed for the trap. Walking slowly up to it, his head swaying and his forked tongue darting constantly in and out, the animal studied the opening which led to the tempting bait.

The hunters, watching breathlessly from the blind, ready to yank the trigger rope, were in an agony of suspense. This was the beast they wanted, no doubt the most monstrous any civilized man had ever seen, and they wanted him alive! Yet always, just as he seemed about to rush in and seize the bait, he would back away and sit staring into the jungle. How long could this go on?

"Just at this moment," says Burden in his account of the expedition, "I heard a vague hum in the distance. It grew louder and louder, and then, in a great roar, something seemed to be descending on our heads, as if an airplane were diving upon us with the engine full on. But it passed over us. The sound of millions of wings

104

filled the air; a great swarm of jungle bees passing low
through the forest. The sound died away again into a
mysterious hum barely audible, and after that I was
conscious of a deathly silence save for a slight rustling
of leaves overhead. The big lizard still remained im-
movable, as though fascinated by a sound he perhaps did
not even hear. Then, all of a sudden, it happened."

The monster, as if the thought had been in his ancient
mind all the time, lunged through the trap opening,
stepped through the hanging noose, and with his great
jaws seized the dead boar. Burden, beside himself with
excitement, tugged at the rope and released the trigger
which held the bent tree. The tree, in turn, held the
open noose that now encircled the dragon's body. With
a mighty whoosh, the tree straightened itself into the
air. The rope grew taut as the noose tightened about
the heavy body and jerked it off its feet.

At last! A gasp of grateful relief sounded in the blind.
But the hunters had not reckoned with the great weight
of the dragon which, falling back to earth, pulled the
tree down again and snapped it as if it had been a
match. Belching violently and emitting a stench which
would have been no more unpleasant to approach had
it been flame that he breathed out, the trapped dragon
struggled to free himself from the heavy rope.

DeFosse, the jungle hunter, stepped forward, carrying
a lasso with which he had been practicing for just such

DeFosse stepped forward, carrying a lasso

an occasion. Carefully he approached the thrashing beast, avoiding the huge tail and the great hooked claws. Several times DeFosse threw the rope, but each time he missed the animal. He calmly recoiled the lasso and tried again.

Seizing a moment when the revolting jaws had snapped shut, the old hunter at last succeeded in slipping the lasso over the great neck. Leaning back, he pulled the noose tight and fastened the other end of the rope to a tree. Then he hurried up to tackle the fiercely lashing tail.

The native boys, now that the struggle was over, came up and helped lash the ugly body to a heavy pole. Groaning and grunting, they lifted the pole to their shoulders and staggered back to camp with three hundred pounds of dragon.

A heavy cage of stout timbers covered with wire cable had been prepared to receive the beast. Into this he was thrust with great difficulty. When the ropes which bound him were cut, the monster showed that he was not yet beaten. He lashed about with such fury that his captors were afraid the cage might not last. The stench of his belchings was so overpowering, however, that no one cared to stay near by. They felt certain that if the beast were not able to break through the cage he would exhaust himself with his struggles and sooner or later settle down to captivity. In the morning they would photograph and measure him.

That is what they believed. In the morning they realized their mistake. When they went to see how the captive had survived the night they found that the wire cable which had covered the breathing hole on top of the cage had been torn as if it were thread. The great dragon of Komodo, whose ancestors had survived the battles and changes of so many millions of years had shown that his time had not yet come.

With increased respect for the strength and resourcefulness of their quarry the members of the expedition got going in earnest. They soon had further evidence of what they had to contend with.

One morning, Mrs. Burden and DeFosse set out to visit the trap in the jungle. It had been reset with another, sturdier tree for a spring pole and baited with the carcass of an entire deer. When they reached the blind opposite the trap, they were astonished at what they saw. The trap had not been sprung, yet the deer had been torn in half and the entire hindquarters— haunches, legs, hoofs, and all—wolfed down in a single gulp. This fact they verified in a very unpleasant manner when their quarry was captured later.

As the monster had gone off into the jungle to digest his 40 or 50 pounds of meat and bone, DeFosse and Mrs. Burden separated and began a cautious search for him. DeFosse went along the jungle trail on the downhill side.

Mrs. Burden walked in the opposite direction, unwisely

leaving her gun in front of the blind. Suddenly, at the edge of the jungle to her right, she saw something move. Sinking back into the deep grass, she watched another great beast make his lumbering way along the trail toward the trap. With a shock of horror she realized that she was standing between the monster and what was left of the bait.

Chilled by the fear that what had happened to the deer might happen to her, she lay shaking. Quickly she tried to decide on her next step. Should she get up and run and so lose one of the largest lizards they had yet seen or should she wait where she was on the chance that DeFosse would return in time?

Let her tell you of her feelings during that awful moment:

Nearer he came and nearer, this shaggy creature, with grim head swinging heavily from side to side. I remembered all the fantastic stories we had heard of these monsters attacking men and horses. Now listening to the short hissing that came like a gust of evil wind, and observing the action of that darting, snake-like tongue, that seemed to sense the very fear that held me, I was affected in a manner not easy to relate.

The creature was less than five yards away, and that subtle reptilian smell was in my nostrils. Too late now to leap from hiding, I closed my eyes and waited.

Then I opened them in time to see DeFosse's head appearing over the hill. The next instant there was a flash, and a bullet buried itself in the great monster's neck. Like lightning he whirled and crashed toward the jungle, but once more the rifle did its work and he lay still!

The Komodo dragon had been able to survive all the chances and changes of many millions of years, but man's thirst for knowledge and his powerful weapons proved too much for the ancient beast. The Governor General of the Netherlands Indies, who controlled the tiny island of Komodo, had given the Burden Expedition permission to capture 15 dragons for scientific purposes. When the time came to leave the thrilling and beautiful spot, the collectors had two living, full-grown specimens safely housed in enormous cages. Their collection also included 12 dead dragons that had been carefully preserved and were ready to be mounted and exhibited in museums.

It was apparent that if man had shared Komodo Island with these fantastic creatures for any length of time they would have become extinct, as they were believed to be before 1912. Nevertheless, when the Burden party left the island in 1926, there were plenty of living specimens remaining. The strange colony of prehistoric beasts is undoubtedly thriving today and if left alone will continue to thrive.

Prehistoric? Yes, the dragon lizard of Komodo is truly prehistoric in that it has survived almost unchanged since what are known as Eocene times. (The term Eocene

comes from the Greek word for dawn, and it is used to describe the period during which mammals—the highest class of animals—began to roam the earth.) You may, of course, wonder of what possible use it can be to spend time and money and risk human life to bring to light a creature which is so far behind the times. What sort of gateway does such an expedition open up? There are many answers to such a question.

Man is interested in his own future, and everything he can learn about the past helps him to guess or imagine what he may expect of the future. The dragon lizards of Komodo had other prehistoric cousins, some of them even larger, but the larger ones—some of which lived in Australia—have not survived. Whatever we can learn about an animal as durable as the Komodo dragon is bound to shed some light on the durability, or lack of it, of other species. Yet perhaps the most important and fascinating things about scientific discovery are the questions which each new find makes us ask.

In the case of our unlovely but interesting dragon, one of the most important questions is: Just what conditions enabled him to survive on Komodo Island? The island is of such comparatively recent origin that it certainly was not in existence when the first of the dragon lizards appeared on earth. There is no evidence that these animals ever lived in a much greater area than they now occupy. They must have done so, however. Where then

112

did they come from? Where were they when the fiery blasts and floods of volcanic lava which caused Komodo to rise from the sea were making it impossible for anything to live between Sumbawa and Timor?

At the American Museum of Natural History in New York you may see a group of three Komodo dragons in a reproduction of their East Indian home. It is like looking into the far past to see them there under the palms with strange volcanic peaks outlined against the sky behind them. You may tremble a little as you watch them and realize that 12,000 miles away, in the jungle of Komodo, that very scene is being reenacted with living creatures. There the great male still lifts his head to watch for the approach of an enemy, while the female tears with her horrible jaws at the dead boar that lies between them.

5. The Arctic: Foe and Friend

The Canadian Arctic Expedition of 1913–1918

Ever since men first appeared on the earth, they have been trying to increase their knowledge of the world and what is in it. Sometimes the path to knowledge has been easy. At other times, false ideas have led men badly astray. The result has been that man's progress toward complete knowledge of his world has been something like the frog's progress out of the well—he has slipped back a step for every two he has moved ahead.

Nowhere has man been more seriously sidetracked by false notions than in his attempt to squeeze the last drop of knowledge out of the polar regions—the Antarctic, or south polar region, and the Arctic, or north polar region.

Let us take a look at the North Pole, which was reached by the American explorer, Rear Admiral Robert E. Peary, on April 6, 1909. The North Pole is, as everyone knows, a position, not a thing—not quite truly a place. It is the name given to the northern end of the axis about which the earth rotates.

Even the ancients of many thousands of years ago knew that, whatever else the North Pole was, it was on ice. What we have learned, only comparatively recently, is that the ice at the top of the world is not fixed. Instead, it is a floating mass, here today and gone tomorrow on an ocean more than two miles deep. If Admiral Peary had remained sitting on the ice at the North Pole he would have found, after a long period of time, that he had moved to another place, for the ice is in motion.

Perhaps we have added to our knowledge of the North Pole since the Admiral's time, but his exploit was the work of a great and courageous man. He got to the Pole because he learned how to travel on ice in the far North, a thing which no explorer before him had learned sufficiently well. That was and will remain a great achievement, although the scientific value of the discovery of the Pole is slight.

The North Pole was not the only thing which took men to the Arctic. There were and are many other things, two of them being especially important. The first was the search that took Columbus across the Atlantic: an attempt to find a sea route from east to west, or, for that matter, from west to east. The second was a desire to find undiscovered land.

No matter what has taken men to the far North in the course of Arctic exploration over many hundreds of years, every explorer has had to deal with one major enemy—the cruel and ever present ice. It is not strange, therefore, that most polar explorers have regarded ice as the only thing to be conquered.

This feeling is one of the things which have affected man's view of polar or, let us say, Arctic exploration. Perpetual ice suggests barrenness just as, to most people, a desert does. And barrenness suggests the absence of life. Men's ideas about how to approach and deal with the Arctic have largely been based on the idea that a trip to the polar regions was somewhat like a long swim under cold water.

If you are going to swim under water, you take a deep breath and hold it. You will be able to stay under the surface only as long as the breath lasts. If you are going to travel in the Arctic, men have thought, you must take with you the means of sustaining life and

117

you can stay alive in the Far North only as long as
your supplies last.

Now if you take a deep breath and dive under water
but get caught on a snag or between two rocks, your
breath will give out and you will drown. Similarly, if
you are caught in the Arctic ice, as practically all explorers using ships have been at one time or another,
your supplies will give out and life will come to an end.

In spite of man's talent for figuring out ways of doing
almost impossible things, many hundreds of lives have
been lost in the North because of the take-a-deep-breath-
and-hold-it attitude toward exploration. This way of
thinking caused one of the greatest tragedies of the
Arctic, but, curiously enough, what resulted from that
tragedy was the development of modern Arctic exploration.

In 1845 two strong ships, the *Erebus* and *Terror* of the
British Navy, set out from England under Sir John
Franklin, carrying 129 men. The purpose of the expedition was to find a passage from the Atlantic to the
Pacific through the ice and across the unknown lands
north of Canada. Sir John was a remarkable officer,
courageous and wise, though not physically strong. His
ships and men were tough and experienced. His supplies
were thought to be enough to last for five years.

In late July of 1845 the two vessels were seen by a
whaling ship at the entrance to Lancaster Sound, some

The vessels were seen at the entrance to Lancaster Sound

800 miles north of the upper end of Hudson Bay and about 900 miles east of the Beaufort Sea. Not a single man of all the 129 in the expedition was ever seen alive again, except perhaps by Eskimos, and neither of the two ships was ever located. A record telling what the expedition had done up to April, 1848 was found by a later expedition, but what happened to those who had survived until then can only be guessed.

For many years thereafter, organized parties searched for Franklin, but they found only scattered traces of his expedition. However, the searchers added more to our knowledge of the Arctic and its ways than men would have learned in 50 years if they had not been driven by the search for a lost hero. As a result Sir John Franklin contributed more than he knew to man's mastery of the polar world.

Yet even the most experienced and skillful explorers were slow to profit from the lessons of disaster and death. It remained for another great leader and another expedition, 68 years later, to show how the Arctic might be conquered and its dangers lessened. Strangely enough the Canadian Arctic Expedition of 1913-1918, led by Vilhjalmur Stefansson, though it brought to an end the era of take-a-deep-breath-and-hold-it ice travel, was also the last great organized ground expedition to the Arctic.

Vilhjalmur Stefansson was a young man of Icelandic descent born in Canada and educated in the United States. Although he was only 34 years old in 1913, he had already spent six years in the Arctic traveling with two different expeditions. He had been trained as an anthropologist—a person whose chief concern is the natural history of man. In his first two trips to the Arctic he had made a study of the Arctic natives, discovering a group of Eskimos who knew almost nothing of white civilization and who were in most ways as primitive as if they were prehistoric beings.

Stefansson had studied the Eskimos not as museum specimens but as human beings who were able to live

120

in a region where white men usually suffered disaster. He studied the Eskimos' diet, their methods of hunting and house building, their clothing, and their means of travel. He decided that if these things enabled the Eskimos to live in the Arctic, they could also help a white man to survive in the polar regions. The more he stayed among the Eskimos, the more he borrowed from them, with the result that he was soon at home in the Arctic, winter or summer, good weather or bad.

Stefansson, however, did not stop with what he could learn from the Eskimos. Not being troubled with some of their primitive beliefs and superstitions, he developed ideas of his own, some of which even the Eskimos said would be the death of him. Having hunted seals and bears on the sea ice north of Alaska, he saw no reason to believe the Eskimos when they said that if you went too far from shore, you would find no seals and bears.

In 1913, Stefansson suggested a trip to the Arctic which would be one of the most completely organized scientific expeditions ever to set out. He planned two sections. The first was to go as far north as possible through or on the ice to see if there was any land there. The second was to do surveying and mapping, to study the natives, the plant and animal life of sea and land as well as the minerals of the region, to take weather observations, and to collect any other useful information.

Because the part of the Arctic in which he wished to

work was largely in or north of Canada, the Canadian
government sponsored the expedition. After much care-
ful preparation Stefansson sailed from Nome, Alaska,
in July, 1913, with one large ship and two small ones.
Aboard them were a large body of men—scientific staff
and crews—and the finest collection of equipment ever
gathered together for Arctic exploration.

The great journey had been planned so carefully that
there was a second choice for everything. If conditions
made one course impossible there was another ready to
take its place.

Sooner than the party had expected, the advantages of
such an arrangement became apparent. The *Alaska* and
the *Mary Sachs*, the two smaller vessels, made their way
eastward close to the shore of northern Alaska and
reached a point where winter quarters could be set up.
The largest of the three ships, the *Karluk*, was not so
fortunate. By the second week in August she was stuck
fast in the ice and never again moved under her own
power. It was not, of course, known at the time that the
Karluk was done for, but it was taken for granted that
she would have to spend the winter in the ice.

This was a blow, but a possibility which had been
considered. It was a blow chiefly because all the members of the expedition had expected to meet at the winter
quarters site where the scientific parties could be as-
sembled and the equipment divided up. As it was, some

of the people who ought to have been ashore were on the *Karluk* and some who were already at winter quarters were separated from their equipment.

One of the most important things was to keep those marooned on the *Karluk* well and happy. Stefansson knew that a plentiful supply of fresh meat would help. As the ship was stuck in the ice near a region which the Commander knew to be good hunting ground, he took three of the expedition staff, two Eskimos, and three sledges and made his way ashore over the ice. He intended to be gone no more than a week or two. One of those who accompanied Stefansson on this trip was a young Australian photographer who later, as Sir Hubert Wilkins, became one of the great names in Arctic and Antarctic history.

While the hunting party was on its way ashore, a terrific storm came up and the men had to camp until it was over. When they again were able to look out to sea, the *Karluk* had disappeared. Though she was later reported from several places along the Alaskan coast, none of the shore party ever saw her again.

Stefansson knew that she might turn up in the spring, but he feared that she might not. Here he was, Commander of the entire expedition, with half his men and supplies and his strongest ship taken away from him. It was a very difficult position to be in.

Stefansson's part of the expedition's work was to have

been the exploration of the Beaufort Sea, that lies between Point Barrow, Alaska, and the Parry Islands, north of Canada. His task was to search for new land both in the Beaufort Sea itself and north of the islands of the Parry Archipelago.

Now, deciding that the loss of the *Karluk* would not justify a winter's idleness, he determined to carry out his plans. He would not waste time worrying about the men marooned on the ship, for they had means of getting ashore over the ice if the vessel were crushed. Furthermore, they had plenty of supplies to tide them over.

Acting upon his decision, Stefansson spent the fall and winter of 1913-1914 in traveling on the Alaskan coast of the Beaufort Sea, picking up Eskimos, dogs, sledges, and supplies and planning the work of both sections of the Expedition. The group of scientists who made up the Southern Section went into winter quarters, supposing that they could do no field work until spring. They were not sympathetic with Stefansson's desire to continue his exploration of the ice of Beaufort Sea. Most of them believed that without a ship he would be able to do nothing. He knew better, and he was ready to travel by March, 1914.

A man standing, at that time, on the shore of northern Alaska looking toward the Pole, would have seen an almost unbroken wilderness of ice under a steely sky

streaked with black patches where open water was reflected against the clouds. But the ice which he would have seen was not the ice of skaters and winter carnivals. It was mile after mile of miniature mountain range, with great ice blocks thrust up as high as houses and the gulleys between them filled with shattered fragments that were covered by soft snow.

It was into this terrifying world that Stefansson and

two companions with one sled and six dogs set out before a howling gale on March 22, 1914. Although they carried supplies which would have fed them for a little more than a month, Stefansson planned to remain on the ice for many months if necessary. He believed that with proper clothing, scientific equipment, cooking utensils, guns, and ammunition, a small party of men could live for any length of time in the Arctic.

There were those in the Southern Section of the Expedition who did not agree with him and did not hesitate to say so. They felt, and expressed the feeling publicly, that the Commander was a little unbalanced. They also believed that he was committing a rather spectacular but useless kind of suicide.

Before starting out, Stefansson asked that the ship *North Star* be sent to meet him at Banks Island, the great land which flanks the eastern side of Beaufort Sea. When he did not return after a few months, his opponents in the Southern Section decided that it would be useless to send the ship. Why risk a good vessel in the ice when the sledge party had already perished and would certainly never be found? It was a simple matter of arithmetic. Stefansson had carried food for 40 days and he had been gone much longer than that. How could he be alive?

Let us take a look out on the ice and see what was happening to the "unbalanced" explorer and his party. When Stefansson had set out from the Alaskan shore,

127

Mile after mile of miniature mountain range

several extra dog sledges and experienced men had accompanied him, carrying additional supplies. On April 7 the Commander sent these parties 50 miles back to shore with letters telling of his plans. The letters also instructed the *North Star* to take aboard a cargo of ammunition, sleds, and scientific equipment, to cruise northward along the Banks Island shore very early in the season, and to look for him among the northern islands.

Stefansson's permanent ice-party then headed northeast into the unknown. It consisted of the leader himself, Storker Storkerson and Ole Andreasen, and a 200-pound sled loaded with 1,200 pounds of equipment and pulled by six dogs.

The first day they made only a few hundred yards before being stopped by open water. On the second day this stretch of water—called a "lead"—closed up, but men and dogs could only go about a mile before another lead forced them to make camp and wait. On the third day they made two miles toward the north but were stopped by snow and high winds which grew fiercer after camp was made.

While the gale howled about the flapping tent, the three explorers lay inside listening to the grinding roar of breaking ice and the crash of huge blocks falling from ridges pushed up by the pressure of the wind. These ridges, sometimes 30 or 40 feet high, kept rising all about the tent. Sometimes they were so close that if a block

They were stopped by open water

had fallen from its moving crest it would have crushed tent, dogs, men, and sled as easily as you could crush a fly.

In the roaring darkness a polar bear passed within 15 feet of the tent and within five feet of the dogs, where his tracks were found in the morning. So terrible was the fury of the storm that neither men nor dogs heard the

huge animal, and he apparently did not know what choice
morsels he had missed.

After the storm, the ice slowly cemented itself together
by freezing. This made travel somewhat less dangerous,
though it was a little like trying to cross a forest by
sledding over the tops of the trees.

As the three men advanced northward they took regu-
lar soundings to test the depth of the sea beneath them.
They found that it was growing deeper and deeper as
they went on, finally becoming too deep to be measured
by the 4,500 feet of wire in their sounding apparatus. On
they went, in an effort to cover as much ice as possible
before spring, with its dangerously warm sun, could
soften the ice beneath their feet. Because of the need for
hurry, the travelers did not start hunting for food until
their sledge load of supplies was fairly well used up.

After a month on the ice, observations showed that
they were about 200 miles at sea and a little northeast
of their starting point. There was no sign of land. The
ocean kept growing deeper. The ice on which they were
traveling seemed to be drifting slightly to the east.

By April 25, Stefansson decided to travel at night and
rest in the daytime. The nights were not entirely dark
and the ice was in better condition when the sun was less
strong. It soon became apparent that the spring was too
far along to allow safe ice-travel to continue indefinitely.
Leads which in March would have frozen over so that

they could be crossed with sled and dogs now, in late April, refused to heal over with anything but slush-ice not safe even to walk upon.

In this place, where no man, white or Eskimo, had ever been before—a place 200 miles from any land—Stefansson had to come to a decision. Should he return as he had come, knowing that the ice was growing worse every day, or should he turn east and strike for Banks Island or Prince Patrick Island, as he had originally planned? His decision was in favor of the latter course. It would take him across a part of the sea through which no ship ever had traveled and which must therefore be unknown and a proper subject for exploration.

This decision, about which Stefansson's two companions were not very happy, meant at least six weeks more on the ice. The food which they had brought with them would only last a few days more and they had, for some time, seen very few seals or bears. On the 7th of May, however, a seal appeared—too far away and on ice that was too soft, but it was a seal. Instead of wasting time in a hopeless attempt to get the seal, the men sat down and ate a huge meal of their fast-disappearing food, knowing that they were back in seal country and would soon need store groceries no longer.

It was more than a week before they were able to get their first seal, but the animal provided fuel, in the form of oil, as well as meat. After that they were never hungry again. In fact, on the 21st of May they had to remain in camp without traveling because they had eaten too much the day before!

Now, as the days became increasingly warm, the leads of open water became wider and more numerous. These had to be crossed by wrapping the sled in a huge water-proofed canvas cloth called a tarpaulin, thus making a sort of boat out of the sled. It could then be used to ferry men, dogs, and equipment over the water. On May 22, when they had been on the ice for two months, the floes began to move westward, away from the land for which the explorers were heading. They were able to travel eastward for a while, thus managing, with considerable effort, to stand still.

On the 24th of May they were completely stopped on a great island of ice 50 feet or more in thickness and four or five miles square. It was surrounded by open water too wide and too rough to cross in the tarpaulin-covered sled.

Stefansson knew that they could ride on this floating island of ice almost indefinitely without fear of its breaking up, but they would have to go where it went until it froze into the sea ice the following winter. That meant that they would have to spend the summer, and possibly the winter, where they were. They would have to store up during the summer enough seals to provide food and fuel

By June 3rd they had killed five bears

through the winter darkness in which no hunting would be possible.

There were plenty of seals about but they were hard to get, and there were, as is usual where there are seals, a quantity of not very friendly polar bears who kept stalking the dogs. By the 3rd of June the men had killed five bears.

On June 5, after their island had taken the men 90

miles out of their course, they were able to cross a lead to the east and ferry 1,000 pounds of lean meat and fat to a smaller floe. The party was now able to travel steadily but slowly southeast—by walking northeast!

On June 22, three months after leaving Alaska, Stefansson looked to the east where he thought he saw the outline of land. The next day the shore was plainly visible. It was not more than 10 miles away! This was not new land, but Norway Island, a small bit of land off the west coast of Banks Island. Although it was not shaped the way the map showed it, it was exactly what they were looking for.

The ice of Norway Island was covered with alternating areas of pool and deep slush through which men waded, dogs swam, and the sled floated. Uncomfortable as it was, this ice was aground and therefore safe. It could not drift away and take the explorers where they did not wish to go. On June 25, after 96 days at sea during which the men had traveled 700 miles (to cover an actual distance of about 500) they stepped ashore.

Here they found grass and flowers in profusion, birds, bees, flies, and the tracks of caribou (the North American reindeer). Stefansson, walking inland, soon killed a number of the animals and set up camp where they had fallen. Caribou are extremely good eating and were a welcome change from seal and bear.

All through the pleasant Arctic summer the three men

hunted caribou, drying meat for later use and curing the skins for winter clothing. They explored the interior of Banks Island and surveyed much of its west coast, always keeping an eye to the south from which direction they expected the *North Star* to come to meet them.

August, the best season for sailing along the coast, came to an end, but still the *North Star* did not arrive. It began to be certain that she never would. Not knowing the truth at the time, Stefansson believed that something must have happened to her. It never occurred to him that she had never been sent.

On September 1 the three troubled men started south along the Banks Island shore toward a point known as Cape Kellett, where there was a possible shelter for ships. Stefansson thought it likely that if there had been trouble a message might have been left at this place or a cache of supplies set up.

By the 11th of September the explorers reached Cape Kellett. They were thoroughly discouraged. Then their feelings changed when a footprint in the silt and the appearance of a ship's mast above the low Cape indicated that all might yet be well. As Stefansson came in sight of the ship he saw that it was not the ship he had ordered to meet him, but the *Mary Sachs*, one of the vessels assigned to the Southern Section. She was drawn up out of the water and her cargo had been unloaded. Near by, several men could be seen building a house.

With a gasp, the man dropped the tool he was holding

As Stefansson approached, he thought that the men saw him but they kept on working, apparently paying no attention. Puzzled, he went closer. Finally one of the men looked up and, with a gasp, dropped the tool he was holding. He stared incredulously. When he realized whom he was seeing he cried out, "Stefansson is alive! He's here!"

To say that the man's surprise puzzled Stefansson would be to put it mildly. Had he not said that he would arrive here at about this time? What, then, was the cause of the strange reaction?

The truth was that since the Southern Section believed Stefansson dead it was logical for the second in command, Dr. Andersen, to take over the leadership. This involved the cancellation of Stefansson's orders and instructions, among which as we know had been an order to send the *North Star* to Banks Island with men, supplies and equipment. Dr. Andersen wanted to keep the *North Star* so he sent the *Sachs* instead, although Stefansson had felt that her twin propellers made her unsuitable for ice work. His feeling was proved correct by the fact that the ship had now lost one of her propellers in the ice; she had sprung such leaks that she had to be pumped for 40 minutes out of every 60; and she could make only two miles an hour. So her crew had hauled her ashore, forgetting that they had no strong timbers on which to slide her back into the water. She was done for. So her crew had decided to build a house as a winter dwelling place, hoping to be picked up later by the Southern Section.

Since the southern party believed Stefansson dead they did not consider it necessary to send the equipment he had asked for. Hubert Wilkins, one of Stefansson's supporters in the expedition, was on board the *Sachs,* but even he did not believe the Commander could have survived. Wilkins had planned to search the coast for traces of Stefansson but without any real hope of finding anything.

Although the *Mary Sachs* had brought a quantity of supplies, they would not be enough to last the entire party of men and dogs for the whole winter. Fresh meat was needed, and it was getting late and the party was short of hunters. Stefansson now turned around and went hunting for them—the rescued feeding his rescuers!

After spending the winter at Cape Kellett preparing food and clothing for the year's work, Stefansson started north again. Once more he used the same methods of travel which the Southern Section of his Expedition still believed (and continued to believe for another year) had led him to his death. It was not until September, 1917 that he and his companions returned to the mainland.

From March, 1914, until that time they were constantly in the field, using what supplies they had as long as they lasted, and then depending on the country for food, fuel, shelter (in the form of snowhouses), and clothing. Between the fall of 1914 and the fall of 1917, they discovered many new lands, among them the islands now called Borden, Brock, Lougheed, Mackenzie King and Meighen; they mapped the coastlines of others which had been incorrectly charted. They were also able to prove that a land which had been on the maps for many years did not really exist.

When Stefansson went north from Cape Kellett in the spring of 1915, he knew that the *Karluk* had been crushed by the ice hundreds of miles to the west. He believed that the members of her crew and staff had made their way safely to shore, using the methods of ice travel which he himself had worked out. He did not know of the tragic and probably avoidable loss of some of the *Karluk's* people who found it hard to cope with conditions which to Stefansson seemed a part of the day's work.

There is in the history of Arctic travel no record of an achievement equal to Stefansson's. If the Canadian Arctic Expedition of 1913-1918 had not come just before the airplane was ready to take its place in Arctic work, Stefansson's methods would have revolutionized Arctic travel. They did prove the gateway to a vast new world of knowledge of the polar lands.

Men never seem, once they get the fogs of superstition and preconceived ideas out of their minds, to lack the means of doing what their forebears have believed could not be done. Even if explorations and discoveries add only a tiny grain at a time to man's knowledge, they still keep showing, as Stefansson did, that the impossible is not beyond man's accomplishment.

Index